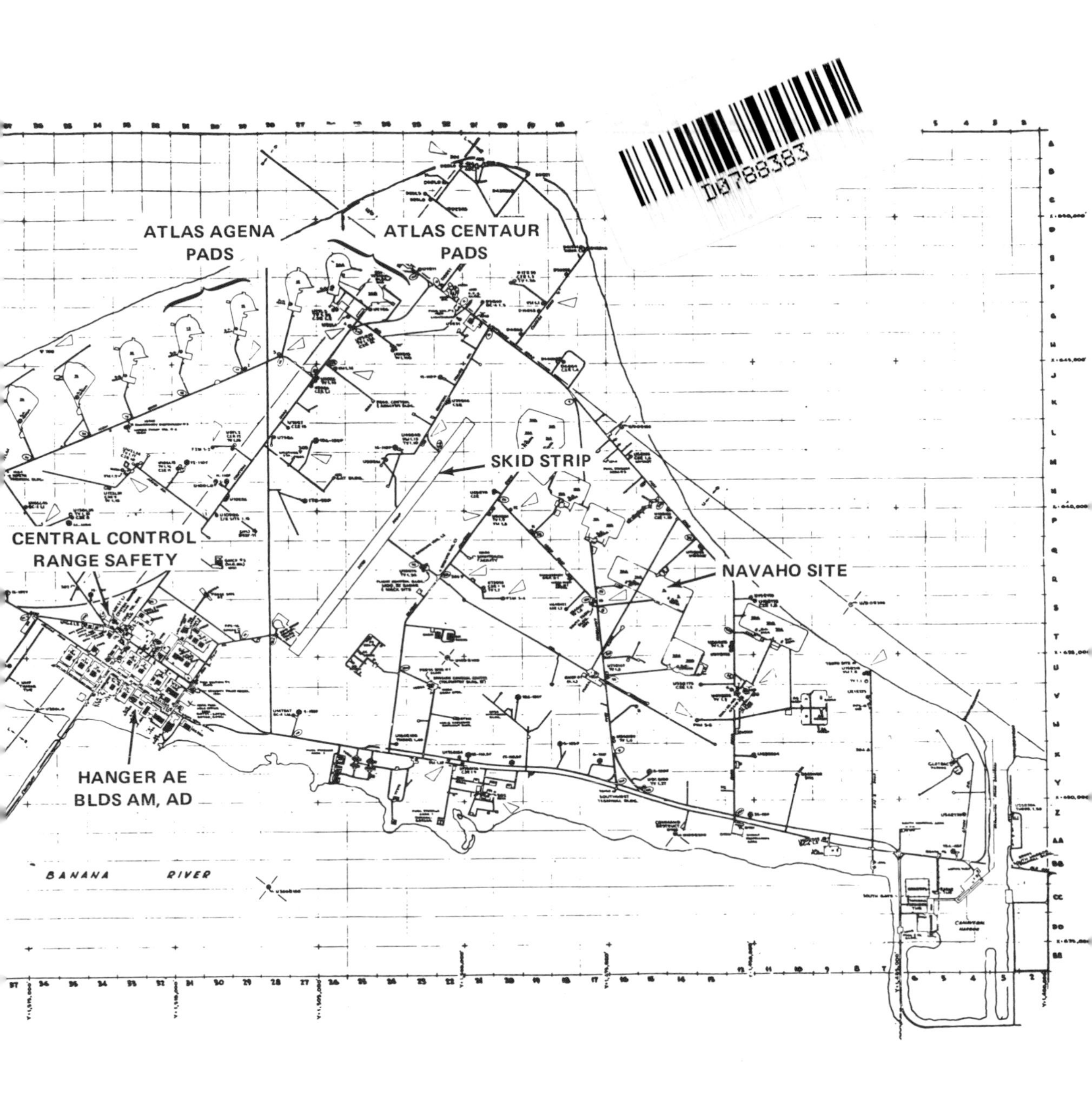
ATLAS AGENA
PADS
ATLAS CENTAUR
PADS
SKID STRIP
CENTRAL CONTROL
RANGE SAFETY
NAVAHO SITE
HANGER AE
BLDS AM, AD
BANANA RIVER
D0788383

FAR TRAVELERS

The Exploring Machines

NASA SP-480

FAR TRAVELERS

The Exploring Machines

Oran W. Nicks

NASA Scientific and Technical Information Branch 1985
National Aeronautics and Space Administration
Washington, DC

Library of Congress Cataloging in Publication Data

Nicks, Oran W.
Far travelers.

(NASA SP ; 480)
1. Space probes. I. Title. II. Series.
TL795.3.N53 1985 629.43′5 85-1794

For sale by the Superintendent of Documents, U.S. Government
Printing Office, Washington, D.C. 20402

Foreword

The first 25 years of space exploration resulted in extraordinary technological achievements and quantum increases in the scientific understanding of our home, the planet Earth, and the solar system in which it resides. Communications, weather, and other Earth observational satellites have affected, directly or indirectly, the lives of most of us. Man has traveled to the Moon, explored its surface, and returned with samples of our nearest celestial neighbor. Unmanned spacecraft have explored our solar system from inside the orbit of Mercury out to the orbit of Pluto.

NASA's program of lunar and planetary exploration with unmanned spacecraft produced a flood of scientific information about the Moon and the planets Mercury, Venus, Mars, Jupiter, and Saturn, as well as the environment of interplanetary space. Some startling discoveries were made: a dust storm that completely covered Mars at the time Mariner 9 went into orbit around the planet cleared to reveal a huge crater and a canyon larger than any on Earth; chemical reactions caused the Viking lander's biological instruments to indicate active results that were not of biological origin; Voyager discovered a ring around Jupiter, active volcanoes on Jupiter's satellite Io, and the strange braided rings of Saturn.

Many challenging engineering problems had to be solved to make these missions possible. Three outstanding accomplishments in this area were: achieving the navigational precision necessary to send Mariner 10 from Venus to Mercury and Voyager 2 from Jupiter to Saturn and Uranus; achieving a soft landing on Mars, whose thin atmosphere required the use of a heat shield for atmosphere entry, a parachute for descent through the atmosphere, and rocket motors for the final touchdown; and transmitting color television pictures of Saturn back to Earth from over 1 billion miles away with only 20 watts of power, the amount used by a refrigerator bulb.

The story of the development of these autonomous exploring machines and the missions they accomplished is one of outstanding engineering and scientific achievements. There were heartbreaking failures, great successes, and some brilliant technological detective work during both the development and flights of the spacecraft. It is also a story of organizations, people, personalities, politics, and outstanding dedication; there are many different perceptions of the relative importance of these factors and the roles they played in the achievements of the golden era of solar system exploration. This is the case not only for people who only observed the program from the outside but also for people who were an integral part of the enterprise.

As a senior NASA official during most of the first quarter century of space exploration, Oran Nicks played a major role in shaping and directing NASA's lunar and planetary programs. His story of this magnificent enterprise provides an important account from one who had great personal commitment and dedication.

H. M. Schurmeier
Jet Propulsion Laboratory
February 1985

Preface

This book about exploring machines is the result of the vision of the late Frank "Red" Rowsome, Jr., Head of NASA's Technical Publications Section. It began as a partnership; I was to provide a manuscript, and he was to make it presentable. Red had many years of technical writing experience, and he specialized in explaining technical subjects simply and clearly. Furthermore, he knew the facts, the people, and incidents we were to write about almost as well as I.

Soon after we began to work on the project Red died, leaving me with unforeseen doubts and decisions. I knew that our plans for sharing with others had been compromised, but by this time I was fired with enthusiasm. After reviewing the guidance Red had given me, and with encouragement from NASA officials, I pledged my best to honor our commitment.

In the planning stages, Red and I had many discussions about the form of the book; however, the final result is necessarily my interpretation. To help you understand what the book is, let me tell you what it is not.

It is not a history. Many of the events discussed are history, but this account is far from complete and coherent. On the other hand, it contains no fiction that I am aware of, and all of the characters, places, and incidents are real—at least as I saw reality. The accounts are largely personal and are therefore limited to my viewpoint or to the views of acquaintances who shared their experiences with me.

It is not a scientific report, although it is an attempt to share some exciting technical aspects of space flight with persons who are keenly interested, including those who have little formal training in technical subjects. I hope the book will also be enjoyed by those who are technically trained, especially those who understand the difficulties of explaining complex space missions and machines.

Although the subject of the book is automated lunar and planetary spacecraft, there are many references to people, and many accounts are written in the first person. It is not, however, an autobiography or a biography of a person or a spacecraft. The people, machines, and incidents are blended in an informal manner, in the hope that the interactive processes involved in creating and deploying spacecraft will be viewed in perspective.

The accounts of the missions, the coverage of the technical subjects, and most of all, the recognition of persons involved, are incomplete. Pangs of conscience stab me often when I see or remember a friend who was overlooked; many would have been worthy subjects for examples not cited. I am sure my colleagues will be reminded of more interesting events I might have used.

Red and I believed that the lunar and planetary spacecraft were the first sophisticates of a new age of machines and that people would want to know more about how they were created, how they worked, and what they did. So many activities occurred in such a few years that memories were becoming blurred, and might soon be erased completely, unless someone tried to write them down. We were fortunate to have been a part of those special events, and we felt obligated to share them, if possible.

I hope you will accept the book as an account of the men, machines, and events as one person saw them, and that you will forgive oversights of deserving people and shortcomings of technical explanations and accounts. The effort will have been worthwhile if enjoyment is realized from revisiting the era when exploring machines reached out to other worlds as peaceful envoys of inquisitive, creative man.

Oran Nicks
January 1985

Acknowledgments

Frank "Red" Rowsome had the idea for this book, and it was also his idea that I write it. He caused me to think about the wonderful but perishable memories from our early space adventures and inspired me to share them with others through this book. He also guided me in the early stages of outlining and researching.

Rebecca Swanson helped do research and typed my initial, dictated drafts before she met her husband-to-be and went his way. William Tsutsui put several draft chapters into better prose, and most importantly, helped me adapt to a new word processor which he mastered quickly. Jo Ann Williamson transcribed many tapes and copied revisions more than once.

I will be forever grateful to Paul Sittler, a "friend" I met for the first time when my word processor failed on Christmas morning, just as I was planning a quiet, productive week of writing during the school break. Paul gave half of his Christmas day to fixing my machine. His unselfish act uplifted me and helped me to write one of the better chapters of the manuscript.

The fun part of writing was reviewing the past with old friends who had been there also. I simply cannot mention them all and the contributions they made without overdoing this greatly, but when they read this, I trust they will know my deepest gratitude is for them.

There were many who read some or all of the manuscript and gave me good suggestions for improvements. T. Keith Glennan honored me most by reading it twice and by offering a number of very constructive comments. Bud Schurmeier provided the very best technical review, and Ed Cortright, Alvin Luedecke, and Cal Broome helped me get on track. A respected writer friend, Jane Mills Smith, and two student friends, Andrew Sullivan and Jon Williams, provided fresh-eyes critiques of real value.

To all of these worthy, respected friends—thanks.

Contents

For Want of a Hyphen

Our friend died violently at 4:26 A.M. on a hot July night. Her finish was spectacular; she was trapped amid the flaming wreckage of an explosion that lit the night sky. Four of us watched helplessly, standing together at a site that gave us a perfect view. We had come there with a common interest in her adventuresome goal, though we came from different backgrounds, and each of us brought a different perspective and commitment to her tragic performance.

We were soon to learn that she had been blown up intentionally by a man with no firsthand knowledge of her ability and promise. My emotion changed from disappointment to bitterness when I learned that she was destroyed barely seconds before flying beyond his reach. We had witnessed the first launch from Cape Canaveral of a spacecraft that was directed toward another planet. The target was Venus, and the spacecraft blown up by a range safety officer was Mariner 1, fated to ride aboard an Atlas/Agena that wobbled astray, potentially endangering shipping lanes and human lives.

Before launch the space vehicle was a breathtaking sight, poised and erect in the night sky, a great gleaming white projectile lit by searchlights so intense that their beams seemed like blue-white guywires. Driving to the launch pad, it was difficult to determine the scale of this bright image in the dark sky. Was it a marvelous Hollywood model, or could it be full sized and real? As we drew closer, it became real, and immense.

I was accompanied that fateful night by Bob Johnson, a NASA protocol officer, who was helping me shepherd Congressmen James Fulton and Joseph Karth to the launch site, slipping us through a roadblock before the area was officially sealed. For a firsthand view of the launch and the disaster that followed, no one had a better position than we four. We stood in the open atop the blockhouse of Mercury Pad 14, less than a mile away from the launch on Pad 12. It was easy to follow the Atlas rocket engines by their firey flame and roar for the 269 seconds—better than 4 minutes—they seemed to operate normally, and even easier to see the tremendous explosion brought about by the destruct command.

Shortly after we had climbed the stairs leading to the top of the blockhouse, guards from the roadblock arrived and asked us to leave. Congressman Fulton was determined to stay. The frustrated guards departed to consult higher authority. I learned later that the Air Force Base Commander was roused from bed to consider the problem. Had time allowed, he would probably have ordered the launch delayed until we were removed, but he was informed too late to intervene.

According to range safety edicts, we shouldn't have been on the roof of the blockhouse. We were there only because of the strong desire and authoritative style of Congressman Jim Fulton, who had already become known as a staunch supporter of the space program. I wasn't particularly worried about our exposure, knowing that range safety requirements were extremely conservative. I would not have chosen that roof as a place to be, with no instruments or communications to give information about the events, but I thought of congressmen as representing the people for whom we all worked and having leadership roles for all that we did. This naive view meant, of course, that we were obligated to do their bidding.

Fulton, a Republican from Pennsylvania and not without an element of theater in his manner, had been in the House of Representatives for a number of years. He was known by the protocol staff at the Cape to have great interest in space, attending almost every launch in the early years. Fulton had the unique habit of collecting souvenir scraps of materials around the pad after a launch; he took the scraps back to Washington and presented them to visitors from his home district.

At that time we did not know our other dignitary, Congressman Joseph Karth of Minnesota, as well. He had only recently been elected to the House of Representatives. As a lawyer and union arbitrator, he had not at first welcomed his assignment to the Space Science Committee, for it had little relevance to his constituents. However, the assignment was a wise one, for Karth later became shrewd and influential in space-related matters in Congress. I attended the launch as Director of Lunar and Planetary Programs for NASA; as senior official present, I had drawn the duty of "babysitting" the congressmen.

We could have watched the launch from many other places at Cape Canaveral. There was the blockhouse, crammed with about 60 engineers and technicians responsible for checking instrumentation showing voltages, temperatures, pressures, and other vital signs from the launch vehicle. Another group of spacecraft engineers in Hangar AE concentrated on detailed instrument readings from the spacecraft itself, the costly and delicate principal actor in the enterprise. The range instrumentation group was in a third building several miles away; they may have had the best "view" of the exact whereabouts and trajectory of the rocket from their elaborate tracking radar displays, but they could not glory in the smoke and flame of a launch, breathtakingly close to where we stood. All these groups were linked by telecommunications, and the spacecraft group was also linked to counterparts at the Jet Propulsion Laboratory, 3000 miles away, working in a makeshift space flight operations facility in Pasadena, California. Tracking stations around the world were in radio contact, receiving basic reports about the countdown and eagerly awaiting the arrival of the spacecraft in their area of the sky. NASA managers and contractors and VIPs were on hand to monitor the launch at the Florida and California installations, as well as at NASA Headquarters in Washington, D.C.

Despite our superb view, we were in the worst position to understand or affect what was happening; we had no communications or messenger, no knowledge of anything except the great fireball high in the sky. Congressman Karth, who had never before attended a launch, asked, "What happened?" Not really knowing, I replied that evidently the vehicle, and probably the spacecraft, had been destroyed, although there was a faint possibility that a clean staging had been achieved before the fireball appeared.

Impelled again by the strong will of Congressman Fulton, we drove to the launch pad, where Fulton began searching for scraps of wire and bits of tape—anything that might have been a product of, or present at, the launch. He filled his pockets and asked us to do the same. With our bits of scrap and gloomy thoughts, we met with project officials at an all-night cafeteria on the base to hear engineers' reports and to compare notes on what had happened. A short time later there was a briefing for reporters; all that could be said—all that was definitely known—was that the launch vehicle had strayed from its course for an unknown reason and had been blown up by a range safety officer doing his prescribed duty.

Engineers who analyzed the telemetry records soon discovered that two separate faults had interacted fatally to do in our friend that disheartening night. The guidance antenna on the Atlas performed poorly, below specifications. When the signal received by the rocket became weak and noisy, the rocket lost its lock on the ground guidance signal that supplied steering commands. The possibility had been foreseen; in the event that radio guidance was lost the internal guidance computer was supposed to reject the spurious signals from the faulty antenna and proceed on its stored program, which would probably have resulted in a successful launch. However, at this point a second fault took effect. Somehow a hyphen had been dropped from the guidance program loaded aboard the computer, allowing the flawed signals to command the rocket to veer left and nose down. The hyphen had been missing on previous successful flights of the Atlas, but that portion of the equation had not been needed since there was no radio guidance failure. Suffice it to say, the first U.S. attempt at interplanetary flight failed for want of a hyphen.

Mariner 1 was not my first exposure to failure during launch, nor was it to be my last. It was also not the first for the Launch Conductor in the blockhouse who was responsible for the Atlas countdown and launch operation. He was Orion Reed, a man whom I had learned to respect during the early 1950s, when we both worked on the Navaho program for North American Aviation Missile Development Division. This Air Force program used rocket boosters for launching ramjet-powered cruise missiles and had provided a base for many of the technologies now being applied to space projects, as well as practical experience for us both.

Returning to Cape Canaveral for a rocket launching after 5 or 6 years was nostalgic. During the Navaho flights my responsibilities had been limited to the ramjet propulsion systems on the cruise missile, and my reason for being at the flight test site was to review the instrumentation checkout for the ramjets. The rocket launchings were to start our missiles on their way with a boost to an altitude of over 60 000 feet and a speed of Mach 3—three times the speed of sound—so that they could begin cruise under their own power. My interest in the launch phase was similar to that for Mariner 1: it was necessary for the booster to succeed before the missile had a chance to complete its mission. One difference was that my overall responsibility for the Mariner program now made me answerable for launch vehicle performance as well as for spacecraft operation.

Reed had been responsible for Navaho flight test operations at the Cape throughout the program; when it ended in 1959, he decided to remain there and joined Convair, the company building and flying Atlas ballistic missiles. All our Mariner, Ranger, and Lunar Orbiter launch vehicles used Atlas boosters, so this put him right in the thick of our early lunar and planetary launches. I was to visit with him frequently during launch operations to come.

Although the Germans had experienced many failures in developing their V-2 rockets, those of us trained in the aircraft industry who had become involved with rocket applications did not have enough respect for rocket development problems to expect or tolerate the failure rate that was experienced. Someone aptly described rocket firings as "controlled explosions." Perhaps that truth, plus the fact that booster operations required far more automation than needed to successfully test fly a new aircraft, were powerful factors.

Critics of the Navaho program dubbed the effort "Project Nevergo"; the project was finally canceled partly because of the booster failures. Since coming to NASA I had been associated with nothing but failures—two Pioneer lunar missions and four Rangers had all failed. Now Mariner 1, after almost 3 years of failures, was a failure too. Mariner had been a special concern to me; it was the first program I was involved in that started from scratch after my arrival at NASA Headquarters early in 1960. I didn't know when or how the failures would end, or if they ever would. My flight from the Cape back to Washington has been erased from memory, but I probably spent it staring out the window with unseeing eyes.

The Team Assembles

Unless you are a sailor, a shepherd, or otherwise occupied outdoors at night, the great sweep of the heavens is almost surely less familiar to you than it was to your ancestors. Hazed, light-polluted skies and indoor occupations mean that the firmament is observed infrequently; when it is observed, it is diminished by the rosy glow from a neon light or a shopping mall aurora.

This was not so for previous generations. In their skies, the ancient constellations silently wheeled like a great stellar clock, marking the hour and the season. The absence of backscattered light and pollution allowed the stars to shine brightly even to a casual observer. Among the changeless, mythic patterns appeared even brighter, uniquely tinted planets moving strangely through the stars, carrying undecipherable messages of fortune, love, and war. The silver Moon, its phases repeating like a stately morality play, long ago acquired associations with hunting, harvesting, love, curiosity, and lunacy. From the time that man became man, the skies have been watched with wonder and awe. As optical instruments became available, men turned them skyward; new concepts and new techniques have always been directed toward the unsolved mysteries of the heavens.

Exploration seems to be in our genes. As they developed the means to do it, men explored the perimeter of the Mediterranean, past the pillars of Hercules, to the sentinel islands off the continent. On land, trudging with extraordinary hardiness through deserts and snow-clogged passes, men traveled as far as their abilities and leadership permitted. If you trace on a globe Xenophon's account of Cyrus' campaigns or the incredible feat of Alexander, who thrust his way from the Mediterranean to India, you must marvel at what armies of men—with cavalry, supply trains, even companies of war elephants—managed to accomplish two and a half millennia ago, under skilled and charismatic leadership. These were no summer campaigns. A

Macedonian trooper who fought with Alexander might have been gone from home for 20 of the most arduous years a man could know.

Nor was simple conquest the sole force behind these forays. On their return, men of Alexander's guard found they had become honored and wealthy citizens, respected in their communities, revered as wise and prudent captains. Many newly subdued lands were a treasure trove, and not the least envied possession of the returned adventurer was his knowledge of exotic lands, river crossings, mountain passes, barbarian tactics, queer foods and languages, and the curious cities and customs of faraway people.

We tend now to think of exploration in a restricted sense—as a scientific, often geographic, expedition, an athletic activity pursued by specialists dressed in fur parkas like Schackleton's or in solar topees like Livingston's. The connotations are overly restrictive if they fail to allow for great tidal movements like the waves of people from Asia that periodically flowed west and south, or for the Scandanavians who crossed the Atlantic in numbers centuries before Columbus. These waves of venturesome people were of a higher order than the random movement of nomads seeking fresh forage. The northern seamen, whose exploits were recorded by poets and genealogists rather than by historians, left us scant knowledge of how they accomplished repeated North Atlantic crossings around 1000 A.D. They must have been skilled sailors to traverse one of the world's most hostile oceans in open boats, making headway against prevailing winds, navigating in precompass days with a primitive latitude technique subject to huge inaccuracies. We must conclude that for some of the species, long and perilous passages were no real deterrent to the exploring imperative.

It is no wonder, then, that the capability of sending instruments into space, and the possibility of venturing ourselves to Earth's nearest "sentinel island," revived the age-old instinct to explore. A handful of intellectual scouts, mainly theoreticians, had examined the idea of space travel. But the dream lay buried until it was aroused by the beep of Sputnik's beacon, which carried all the sudden urgency of a firebell in the night. So imperatively did it sound that the United States awoke from its trance and, in little more than a decade, became a contender for world leadership in space. The events of those days, and the people who made them happen, are our concern.

When NASA was established in 1958, almost precisely a year after Sputnik's signal, it was assembled from diverse organizations. At the nucleus was the National Advisory Committee for Aeronautics (NACA), a middle-aged organization with three highly competent research laboratories. To this nucleus were added complements from the Naval Research Laboratory, the Air Force Missile Systems Division, and, somewhat later, an Army Ballistic Missiles group and the Jet Propulsion Laboratory, all government entities with experience in space-related technologies.

Looking back on the formation of NASA, I recall many discussions and articles that were written about how our nation might muster the leadership and develop the capabilities needed to expand into the new frontiers of space. Their obvious association with missiles and rockets made it seem logical that the military organizations of the Air Force, the Army, and the Navy should be involved, along with their industrial partners, who had been designing and building weapons. One worry of several of us who had worked as military contractors was due to the transient nature of management assignments in the military and to the somewhat unpredictable nature of their policy making where technical matters were concerned. A more significant worry to national leaders was the implication of involving military organizations in the development of the space frontier. To the surprise of many, the merits of establishing space activities as a purely civil venture led to the development of an organizational entity and a program to pursue peaceful space activities "for the benefit of all mankind."

The announced plan to build the new organization around NACA evoked mixed reactions. Many military leaders and supporters were bitter and predicted poor results from assigning leadership and management responsibilities to a relatively unsung research organization. Even though the aerospace industry held NACA in high regard for its research contributions, many felt that the shy, unassuming image that had been a trademark of NACA would not lead to the bold, aggressive programs thought to be needed. Thus, when the announcements were made that gave the new NASA clear control over areas that had been dominated by the military, many believed it would be only a short time before a power struggle would result in the reaffirmation of military leadership.

Fortunately, the NACA heritage proved right for the time. The impressive collection of 8000 dedicated scientists, engineers, and administrators had been working effectively for years on matters closely related to the challenges of space. Knowledge and tools with which to begin the tasks of planning and implementing space activities that would establish the United States as world leader were ready. Perhaps one of the greatest qualities of NACA that could not have been thoroughly appraised in advance of this challenge was a direct by-product of a successful research organization—a certain humility that recognized the need for gathering skills from industry and military circles, without relinquishing leadership responsibilities. NACA researchers were accustomed to searching out contributions that had been made by others and building on them; NASA administrators set about the development of the new organization in the same way. The process was facilitated by an unusual section in the Space Act that gave the NASA Administrator discretionary authority in selecting high-level personnel without the constraints imposed by Civil Service appointment processes. The hiring of 260 "excepted position" scientific, engineering, and administrative personnel began immediately, to complement the NACA transferees and other government employees reassigned from defense organizations.

The Langley, Lewis, and Ames centers had a remarkable array of talent and facilities that were already working on the fringes of space when Sputnik flew. In conjunction with research on high-speed flight, Langley had field operations at Edwards Dry Lake in California and an aggressive rocket launch program in full swing at Wallops Island, Virginia. These field activities were supported by groups who had developed instrumentation, tracking, and data acquisition capabilities that were immediately brought into play. At the time NASA was officially formed, almost 3000 rocket launchings had been made from the beach at Wallops, and about twice that many from research aircraft. Even though many of these rockets were small compared with those needed for launch into space, the vagaries of rockets and the requirements for tracking, telemetry, and data processing were all basic enough to provide a wealth of experience that would be brought into play.

Sounding rockets used to conduct aerodynamic research had become multistage vehicles long before orbital flights were made. Robert R. Gilruth headed a team conducting flight research that included many of the principals who later established the Manned Spacecraft Center in Houston. Edmond C. Buckley, by then Chief of the Instrument Research Division at Langley, was to become the new Associate Administrator for Tracking and Data Acquisition at NASA Headquarters. Robert L. Krieger, who had been involved in the management of activities at Wallops from the very beginning, was named Director of the Wallops Flight Center and remained to continue his lead role for 20 more years. Research engineers, including Clifford Nelson, would become key figures in the Lunar Orbiter and Viking projects.

Being primarily dedicated to propulsion activities, the Lewis center harbored much talent for launch vehicle development and operations. Jet propulsion had been oriented more toward high-speed aircraft than toward missiles, but the principles were the same and many researchers had experience directly in line to advances in rocket propulsion. A noteworthy example is the basic effort that had already been underway on the uses of hydrogen as a fuel. Many studies concerning practical aspects of handling hydrogen, plus a healthy respect for its potential, prepared Lewis personnel for a significant role in space after the opening gun had sounded. They also had a premier knowledge of pumps, seals, high-pressure machinery, and materials that could be called on for immediate advances in rocketry.

Not all the changes from aeronautical research to space technologies had come easily, for good researchers possess a dedication to a line of effort that persists through thick and thin. It was often the foresight of leaders and their ability to redirect researchers to promising new fields that brought changes. John Sloop once told about being reassigned from research on spark plug fouling to beginning efforts at Lewis on rocket research. He knew the problem of plug fouling to be important (it still is, for reciprocating engines), and he thought his research was about to produce a breakthrough. Even though he obediently began to work in the new area on official duty time, he continued to work after hours on his own until he was finally ordered to stop for the sake of his health. His work with hydrogen-fueled engines led to major research results, and in 1960 he was assigned to NASA Headquarters, where he performed a key leadership role in advanced research and technology activities until his retirement. His most recent contribution is an authoritative book documenting research and development activities on liquid hydrogen as a propulsion fuel during the critical period from 1945 to 1959.

The efforts of the Vanguard program to launch a minimal satellite for the International Geophysical Year and the exhortations of zealots like Wernher von Braun, at work on Army missiles, had evoked similar stirrings at the Jet Propulsion Laboratory. JPL, an offshoot of the California Institute of Technology, had developed the JATO rocket concept during World War II to aid aircraft takeoffs, and had later done pioneering work on the guidance and control of tactical missiles. Under the leadership of William H. Pickering, who became Director in 1954, JPL became a national center of excellence in electronics and control technologies, and its scientists and engineers set their hearts on the Moon and the planets.

Bill Pickering had come to the United States from New Zealand. He received a degree in electrical engineering and a Ph.D. in physics from the California Institute of Technology. He taught electrical engineering at

CalTech during World War II and later moved to JPL to work on telemetry and instrumentation for missiles. At the time of Sputnik he began promoting the idea of lunar missions as a logical step beyond simpler Earth orbit flights.

Robert J. Parks, destined to become a key figure in the space flight program, was hired by JPL in 1947, following Bill Pickering's suggestion that the area of guidance and control could become an important part of JPL's future and that they should have an expert in the field. Getting the rockets to fly, although quite an accomplishment, was clearly not enough. Flight control, tracking, and data acquisition methods had to be developed. The guidance and control systems used at JPL, although considered state of the art at the time, now seem almost quaint. Bob Parks recalls that he monitored a contract let by JPL to the Sperry Corporation to adapt aircraft autopilot equipment for use on missiles. Tests were conducted using pneumatically driven gyros with pneumatic pickoffs, amplifiers, and servos on what would become known as the Corporal missile. This direct modification of aircraft-type hardware required the storage of high-pressure gas and did not prove to be a good solution for missiles. Parks eventually became JPL's lead man on planetary programs; in addition to his regular, quite demanding, assignments, he has been called on to rescue faltering projects, and has done so on numerous occasions with great success.

T. Keith Glennan, who had served for about 19 years as President of Case Western Reserve University, was appointed the first Administrator of NASA by President Dwight D. Eisenhower. Hugh L. Dryden, long a wise and honored NACA leader, was named Deputy Administrator. As the years revealed, both were happy choices. Together these men shaped an organization that the United States, indeed the world, learned to respect.

Perhaps because Glennan came from Cleveland and already knew the competent people at the NACA Lewis Research Center (and surely with the concurrence of Dryden), Abe Silverstein, the Associate Director of the NACA Lewis facility, was chosen to be the first Director for Space Flight Programs. This, too, was a fortunate selection, because in the few years he spent at NASA Headquarters, Silverstein played a dominant role in forging the programs and practices and assigning the people that have guided NASA from the beginning. Abe had—and still has—some unusual qualities that never fail to impress (or bewilder and alarm) those who come in contact with him.

Watching Abe deal with presenters of technical briefings, I was often reminded of a story my grandfather had told me about encounters between armadillos and smart (or dumb) hound dogs in Texas ranch country. Even his experienced hound dog, cocky from successful confrontations with coons and skunks, was baffled the first time or two he ran up against an armor-plated armadillo, which would retract into its shell and present a smooth, hard surface too large to bite. All a dumb hound dog could do was to bark his frustration. But this smart, experimentally minded hound discovered that if he flipped the armadillo on its back, there were chinks in the underbelly armor that allowed him to make short work of the miscreant.

On several occasions in the 1950s I briefed Abe and others during wind tunnel tests at Lewis (I was with North American Aviation at the time). It didn't take me long to learn to respect Abe's uncanny insight and unusual style. Fortunately, I was able to answer his penetrating, sometimes intimidating questions without being flipped on my back, but over the years I have seen Abe flip unwary or unprepared briefers and mercilessly rip them open; it was sometimes the only chance they ever got. Abe's was a style that could make enemies, especially of the careless and slovenly, but all who came to know and respect his quick and forthright judgments gave him a lifetime of loyalty.

Another NACA researcher I learned to like and respect during the same period was an engineer, Edgar M. Cortright. Ed was conducting research at Lewis on supersonic inlets and nozzles similar to those we were developing for Navaho propulsion systems. During my visits to Lewis we became acquainted professionally through discussions of related work.

Abe Silverstein brought Ed Cortright to Washington soon after NASA was chartered, to become a principal member of the Headquarters staff. He was involved in developing the series of weather satellites that included Tiros and Nimbus and what would become the synchronous-orbit class of satellites. Ed was also responsible for lunar and planetary programs, which at the time centered around Pioneer and Able missions that had been started under the Advanced Research Projects Agency (ARPA). His first new lunar project, called Ranger, was to become the reason for our fateful reunion and a new career for me.

My contacts with Ed and Abe began again in 1959 when I was working for the Chance Vought Corporation as an advanced projects engineer. During company-sponsored studies on a four-stage rocket vehicle designed for launching small scientific payloads, I made several visits to Washington to discuss proposals and to integrate NASA requirements into our planning. These visits occurred at a time when expansion of the NASA Headquarters staff was underway, and Ed asked me to join the NASA team as head of lunar flight systems, which I did in March 1960. We were to work closely during the next 20 years, certainly the most memorable of my life.

Important decisions were being made in those early years involving matters at a higher level than programs and plans. Glennan and Dryden made a first-rate team, complementing each other almost perfectly, and both were fully aware of the research process. Dryden had personally made many fundamental aeronautical contributions and had for many years studied the impact of research on society. Glennan was the epitome of a judicious, prudent, and skilled senior executive. Becoming leader of a new organization with a once-in-history challenge, and assembling a team of several powerful yet diverse groups was no easy task. Judging the true qualities of people was one of Glennan's greatest assets as a strong leader. Years later he confided to me that he was always privately skeptical of Wernher von Braun's glowing presentations, but von Braun's giant launch vehicles *always* worked.

It is my view that Glennan and Dryden are to be credited with much of the Constitution-like wisdom written into the so-called Space Act of 1958, with its clear assertion that U.S. activities in space be conducted openly and that their results benefit all mankind. Although openness became a hallmark of NASA programs, outsiders may not realize how close we came to going the other way. As most technologies had evolved from missile developments, industry and military officials were accustomed to strict security classifications. Making new technical knowledge widely available was a startling idea, not immediately congenial to defense/industry representatives. But our administrators, after dealing with both classified and unclassified activities for many years, concluded that openness was fundamentally important to scientific advances and to peaceful uses of space. Hugh Dryden told me that he thought scientific openness would be worth far more toward long-term progress than the perishable, uncertain benefits of security that might be achieved by short-term containment. At the time, this position was about as easy to defend as the Ten Commandments; however, I am convinced of its merit today.

The administrative marriage of NASA and JPL in 1959 provided JPL with its long-dreamt-of opportunity to explore the planets, but not without some trauma. JPL had successfully participated in the launch of Explorer 1, the first U.S. satellite to achieve orbit. Under contract to the Army and the sponsorship of ARPA, JPL was also working on Pioneer flights to the vicinity of the Moon. Thus, it was no wonder that Bill Pickering and his staff felt that

JPL had come into the NASA family as a partner, not simply as a contractor, nor even as an analog of the NASA research and flight centers. That ambiguity, combined with a strong esprit de corps and a sometimes prickly pride, caused JPL personnel to resent some of the managerial and organizational directions that NASA administrators deemed appropriate. In spite of JPL's enviable string of successes in space, there are some who believe that JPL efforts would have paid off earlier had its leaders been more willing to accept team assignments, recognizing NASA personnel in their lead roles instead of as competitors.

It was into this environment that I came, assigned a principal responsibility at NASA Headquarters for dealing with JPL. Fortunately, I had become acquainted with several members of the staff while conducting tests at JPL's supersonic wind tunnels for the same Navaho missile program that had also taken me to Lewis. Harris M. "Bud" Schurmeier, Frank Goddard, and others I had worked with were now key figures in space activities and remained staunch allies and friends throughout the years.

In the last year of the Eisenhower administration, NASA's leaders indicated that preliminary planning for manned space flights, active work on communications and meteorological satellites, and Ranger missions to the Moon represented a balanced portfolio of sufficient breadth. When, as a "new boy" in the office, I asked Ed Cortright why NASA had no planetary plans beyond Pioneer 5, he told me that the planets were excluded for the present, until activities already begun were moving toward success. My job at the time had nothing to do with the planets; such missions were assigned to Fred Kochendorfer, another Lewis transferee who was to become Mariner Program Manager. Still, I believed we should be planning planetary exploration in support of a well-rounded space program.

To those of us with an eye on the planets, policy wasn't the only problem. We simply did not have a launch vehicle for planetary missions. To achieve the high velocities needed for Earth escape and planetary trajectories would require multistage vehicles that did not exist at the time. An Atlas/Centaur combination might do, but the Centaur stage, with its high-efficiency hydrogen-oxygen engine, wasn't far enough along for anyone to be sure when, or even if it would come into useful being. Still, a few at Headquarters and many kindred souls at JPL felt that the space program was incomplete with no planetary missions in preparation.

Our opportunity came a while later as a result of new policies announced by the launch vehicle program office. Launch failures had been demoralizing, and as a way of ensuring that vehicle development was complete before commitment to expensive and conspicuous payloads, officials succeeded in convincing the NASA administration that it made sense to devote the first 10 launches to development purposes. von Braun was a persuasive proponent of this approach, and Saturn development proceeded on this basis, with test launches carrying dummy payloads of water and sand.

One argument we offered for planning piggyback planetary missions on Centaur vehicle development flights was that evolutionary development experience was important for spacecraft too; furthermore, both spacecraft and vehicle engineers needed experience in integrating spacecraft to launch vehicles. Both benefits would come relatively cheaply by piggybacking on the Centaur development flights, since the vehicles needed the mass of a payload, dummy or real, for a proper test.

Conditional approval was given for our proposal, and planning began in earnest for planetary missions. It was possible to launch a vehicle to Venus only at 19-month intervals and to Mars every 26 months if we used minimum-energy trajectories, which were all we would be capable of for years to come. Long-range predictions of exact opportunities were made; these were truly firm dates, immune to tampering for the convenience of politicians or administrators. While this immutable quality was an extra challenge to the development problems, it was also a blessing that relieved project planners of the need to justify a particular target date for scheduling and budgeting.

As a condition of hitchhiking on vehicle test launches, we agreed that our spacecraft might be launched at any time and in any direction that suited the launch vehicle test requirements. Thus, flights would not necessarily occur when the planetary launch window was open. We willingly agreed to this condition, not unaware that if we produced a spacecraft on schedule for its mission, the launch vehicle test would be geared to the planetary opportunity if at all possible.

Thus, the Mariner program got its start. By late 1960, plans began to take shape for matching two Centaur test launches with a Venus launch opportunity in August 1962. A Mars opportunity would occur a few months later; therefore, as many as four planetary launches on test flights were possible in 1962.

The infrequent launch opportunities made production-line spacecraft manufacture desirable, so that two spacecraft could be launched at either opportunity. Design studies suggested that for trips inward toward Venus and outward toward Mars, a somewhat standardized spacecraft "bus" would serve if it were fitted with adaptations of standard solar arrays, antennas, and the like. A NASA requirement plan was established, and JPL studies were begun for the multipurpose Mariner spacecraft, designated Mariner A and Mariner B, with A planned for Venus and B for Mars. The size of the Mariners was determined by the Centaur capability, and each had a gross weight on the order of 1250 pounds, depending on the mission energy requirements. The first flights of these spacecraft were planned for Centaur development launches 7 and 8, so there was hope that vehicle "infant mortalities" could be avoided.

During the spring of 1961 we lost confidence that the launches would take place the following year. Centaur was rumored to be in trouble: not on schedule and perhaps not even feasible. Part of the difficulty was that Centaur was linked to Saturn rocket developments at Marshall Space Flight Center. Like Saturn upper stages, Centaurs were to use hydrogen-oxygen propellants. However, Centaur requirements were entirely different from those of Saturn; Centaur was intended for military missions, for synchronous-orbit communication satellite missions, and for planetary missions. Furthermore, the transfer of Centaur liquid hydrogen development responsibilities from ARPA to NASA was affected by the requirements for committee coordination and jurisdictional wrangling. The combination of these factors made Centaur very problematic.

Through contacts with Donald Heaton, Centaur vehicle manager, I received inklings that Centaur was in deep trouble, so deep that the Venus launches planned for August 1962 were threatened. A short time later, on a visit to JPL concerning the Ranger program, I talked with Dan Schneiderman, who had been involved in planetary spacecraft design studies when the Vega stage was being developed. I found my way to his small basement office and discussed the possibility of using an Atlas/Agena to launch a planetary spacecraft. At first Dan was highly skeptical that the 400-pound payload Agenas could carry would be at all adequate for a mission. But Dan had a wonderful knack of thinking positively about a challenge, a characteristic I was to see at work many times in the years to follow. He examined the results of previous studies and concluded that it might be possible to build a spacecraft for an Atlas/Agena launch that could carry perhaps 20 pounds of scientific experiments to Venus.

Armed with this information, I returned to Washington. Within a few days a meeting between the Administrator and the Director of Launch Vehicle Programs took place, and, because our programs depended so heavily on Centaur, Abe Silverstein, Ed Cortright, and I attended. The meeting produced a formal position by Wernher von Braun that Centaur's future was totally uncertain. We left the meeting with the clear understanding that our Centaur-based planetary missions were postponed indefinitely.

While walking down the hall after the meeting I mentioned to Abe Silverstein the results of my chat with Schneiderman and asked whether it was possible to consider flying a modified Ranger on an Atlas/Agena during the 1962 Venus opportunity. He thought for a moment and then said, "I guess Glennan said that we won't be doing planetary launches on Centaur—he didn't mention flying Agena." I knew this was Abe's way of telling me to go ahead without formally giving me authority to do so, but I felt comfortable with this degree of approval from him. I immediately called the JPL people to tell them of the indefinite Centaur delay and to encourage rapid preparation of a plan for a substitute Venus mission using Ranger hardware.

Of course I was helped by the knowledge that JPL was aching to begin planetary missions and that a minimal payload could be carried. JPL snapped at the opportunity, appointed a high-caliber team (Jack James was the Project Manager and Dan Schneiderman was the Spacecraft System Manager), and had a proposal outline ready on August 28, 1961. The Mariner R, so called because it was made from Ranger hardware, would have a high probability of a single launch in August 1962 and a possible second launch if all went well. While this would affect the Ranger schedule, delaying it slightly, the proposal included suggestions on how this could be done without major compromises. Since Cortright and Silverstein had already informally approved the idea, it was with record-setting swiftness that NASA gave formal approval to JPL in early September 1961 for two Mariner R launches to Venus in July-August 1962.

In the 11 months that remained before the launch window opened, JPL had to design, build, test, and integrate two spacecraft for an entirely unprecedented mission. It also had to develop the complete tracking, data acquisition, and operations capabilities needed for a long-term, deep-space mission. So innocently hare-brained an effort would not be approved today, and experienced planners probably would not propose it. From 3 to 5 years would be needed, assuming that parts of the system had flown before. (It now takes 5 years to do almost anything.) Not knowing that the proposed mission was almost impossible, we laid out a plan, reprogrammed funding and hardware, and went ahead and did it.

Creating an Exploring Machine

A great deal of engineering is based on previous work. Improvements resulting in lower costs and greater durability are made, weaknesses are corrected, and limitations are reduced. But what if prior experience is absolutely zero? How do we design, build, and test something that has never been built before? That was the daunting challenge that faced the designers of the first Mariners.

To begin with a blank sheet of paper and attempt to create an interplanetary robot is to confront unexpected problems. Simple solutions may be found for problems that at first appear insoluble, but sometimes deceptively simple obstacles are almost insurmountable. Nature has provided countless living creatures with effective solutions to problems of stabilization and mobility, but endowing a robot with these attributes is not easy. Of course, we do not yet know fully how nature's creatures perform many of their functions. Who can say how a migratory bird navigates for thousands of miles or how a hawk stabilizes its head while turning its body? Perhaps the technical approaches engineers use will seem less complex when nature's methods are finally unscrambled. Considering the fact that it was designed to travel 180 million miles to another planet and make observations, the first Mariner was an extremely simple exploring machine, rudimentary and primitive compared with a human being and even with the spacecraft we would launch only a decade later.

Simply put, Mariner was a machine used by man to extend his powers of observation beyond the immediate vicinity and out into space. Lacking the experience and resources to launch an astronaut deep into the solar system, we sent exploring machines as our proxies. Like the specialized robots in a nuclear facility, interplanetary probes were developed to do a job that human beings were unable to do. In the case of Mariner, the assignment was to perform preliminary exploration of a neighboring planet and to learn as much as possible during the journey.

It may not be surprising, therefore, that exploring machines came to resemble living creatures in a number of ways. Designed de novo using engineering principles and technology, Mariner had remarkably human attributes in performance and in its ability to cope with the environment. While dependent on a rocket-propelled launch vehicle for basic transportation, Mariner had to be capable of acting for itself on the way to its destination. Incoming solar energy had to be assimilated to sustain it. Attitude orientation was required to obtain power, to maintain communications, for pointing sensors, and for thermal control. A method of knowing where it was and where it was headed was required; thrusters were needed to serve as "muscles" for attitude and course corrections. It had to have some memory and a time sense, plus an ability to interpret and act on commands and to communicate its state of health and its findings.

A spacecraft, of course, must be held together by a rigid structure, just as a human body is defined by a skeleton. Refined in design yet simple in appearance, the basic structure of Mariner was hexagonal; it was made of magnesium with an aluminum superstructure. Weight is always a primary constraint in vehicles destined to be launched into space; Mariner's frame was as light as possible, since inert parts competed for the same precious weight that might be allocated to sensors, data processors, and the like. The craft was six sided because of its Ranger ancestry; the Ranger was six sided in part to allow efficient structural attachments to the Vega upper stage and in part to allow convenient mounting of solar panels, electronics boxes, and the midcourse propulsion system. On the superstructure were placed antennas, scientific instruments, and other components needing a location with a vantage.

The electronics compartments and the subsystem compartments around the base were modularized so that they could be separated more or less by function, allowing the development of power, guidance and control, instrument signal conditioning, and communications systems in individual laboratories before these bays were brought together and integrated to become a spacecraft. The six boxes were rectangular so that electronic components could be easily packaged in them, yet they all interconnected around the structure, becoming what is called the spacecraft bus, sharing common power systems, thermal control systems, and other basics essential to component integration.

The structure attached to the top of the bus served many functions, but perhaps the most important was that it carried a low-gain antenna at the upper end. The low-gain, or omnidirectional, antenna provided the primary source of command and backup transmission capability, so that signals could be received or engineering data transmitted regardless of the attitude of the spacecraft. The antenna had little or no amplification of signal in any direction, but produced a radiation pattern similar in all directions. Actually this antenna could not be perfectly omnidirectional; some shadowing by the spacecraft bus in some directions was expected. But, located away from the spacecraft, it did have a generally good "view." In addition, since omnidirectional antennas do not have to be very large or very heavy, they fit nicely in the point of the aerodynamic shroud without unduly affecting the center of gravity.

Other components mounted on the superstructure because of the view advantage included scan platforms that could "see" a planet as the spacecraft went by, or, in the case of the Ranger craft to the Moon, cameras that looked out and down as the spacecraft approached the surface. Components that fitted into the superstructure were like the bus compartments, essentially modularized so that they could be assembled and tested in the laboratory before integration with the spacecraft.

Like its human counterpart, a spacecraft needs a regular supply of energy. The stored power of batteries is one alternative, but for missions lasting weeks or months, some means of replenishing battery power is necessary. Mariner's prime energy source was the Sun, which supplied about 150 watts of electricity through solar cells that could charge an internal battery having a storage capacity of 1000 watt-hours. As long as the solar cells were facing the Sun, Mariner had power to lead its own life in its own way. Even if the panels were shaded, automatic switching systems allowed the spacecraft to operate on battery reserves for a time. Lest Mariner grow too independent, however, there were also circuits that could be commanded from Earth at the discretion of the spacecraft's terrestrial masters.

Early solar cells were fairly inefficient at converting solar energy to electrical energy; only 7 to 10 percent of every unit of energy the Sun beamed onto the cells was converted to useful electrical energy. Nevertheless, the Sun offers a clean, dependable source of energy in space. An attractive feature of solar cells is that they are passive devices with no moving parts that wear out. They do have shortcomings, however. In addition to requiring orientation so that they receive full and direct sunlight, they are temperature and radiation sensitive. Solar panels tend to overheat, so a good deal of engineering work is required to develop the proper thermal environment. Fortunately, the backs of panels facing the black sky can be used to allow heat to escape; by judicious engineering, panel temperatures similar to normal room temperatures on Earth can be maintained.

Other worries by designers of solar panels included harmful effects of radiation and micrometeorite impacts. After some bad experiences with early satellites, solar cells were made less radiation sensitive through the use of better materials and protective covers. Micrometeorite protection was limited to wiring cells so that only localized losses of cells would result in case of hits. Some failures are thought to have been due to micrometeorite hits, but the evidence is inconclusive.

An attractive aspect of solar energy, its constancy, also became something of a challenge to spacecraft designers because the usage rate or demands of the spacecraft varied considerably: there were periods when the requirements might exceed the incoming supply and times when excess power would have to be dissipated. This called for an innate capability to adjust the dissipation of energy when the spacecraft requirements were exceeded by the supply.

Thus, power management involved circuitry connecting the solar panels to the batteries in a semiautomatic manner, for it was not logical to try to monitor and control power usage from Earth. However, in emergencies, commands from Earth to adjust power usage were needed, so both data readouts and command functions had to be integrated into automated power system designs.

Finally, the solar panels had to be folded inside the heat shield or nose cone that protected the spacecraft from aerodynamic forces and from the heating that occurred during launch through the atmosphere. This mechanical consideration, involving latching mechanisms, deployment commands, and dynamics of actuation, produced additional headaches for engineers. In a disproportionate way, the success of the sophisticated solar energy collection and conversion system was totally dependent on simple pyrotechnic and mechanical latching systems, for if the panels did not open and were not exposed to sunlight, the consequences would be disastrous.

A human without attitude control would be sadly handicapped, unable to swing a bat, throw a ball, propel himself, or even turn his eyes away from the glaring Sun. Mariner needed attitude control for precisely the same reasons. It must be remembered that inertial space is a most peculiar place, at least by terrestrial standards. There is no up and no down, no day and no night, no air and no true wind. The Sun and other stars are visible at the same time, and surfaces facing the Sun grow very hot while those in shadow grow very cold. Most remarkably, objects that are moving or rotating continue to move or rotate indefinitely until they are stopped by countervailing forces. A spacecraft injected on an interplanetary trajectory in this odd environment would, lacking a stabilizing system, tumble at random, preserving the last impulse imparted to it, plus the resultants of additional impulses that might be derived from particle impacts or from reaction to onboard movements. Naturally, this kind of random movement will not do if an antenna must be pointed precisely, if scientific instruments are to scan a planetary swath, or if an onboard rocket must be aimed carefully to correct an imperfect trajectory.

A simple way to hold a spacecraft fixed in inertial space is to spin it like a top. The whole vehicle then becomes the rotor in a gyro, holding its polar axis in relation to the orbital plane it traverses. This principle has been used with great success for some Earth satellites and the Pioneer class of interplanetary craft, for which simplicity and long life are important considerations. However, the disadvantages are considerable: a spinning scientific platform, an antenna that must be aligned with the polar axis, and the need to mount solar cells in a drum-like configuration so that spinning won't materially affect power generation. The gyro principle can be applied to a reference platform for sensing attitude and maintaining control, but gyros alone would not do the job reliably over the months-long periods needed for even short interplanetary trips; the best of them would be susceptible to drift arising from the accumulation of infinitesimal errors.

An alternative to gyros is an automatic system to hold the entire spacecraft in an established attitude by sighting on distant celestial objects. A principal in the development of guidance and control systems for unmanned spacecraft was John R. Scull, who continued to be involved through all the lunar and planetary missions of the 1960s and 1970s. He and his associates worked out the application of optical sensors and gyroscopes that became standard for spacecraft guidance and control. This type of three-axis stabilization worked fairly well for Mariner and was improved for later missions. The principles are simple: sighting on distant celestial objects, ingenious sensors keep an instrumental eye on distant "spacemarks." If any substantial straying of attitude is detected, the sensors send signals to paired attitude jets. Each jet is a tiny minirocket that releases a spurt of compressed gas to nudge the spacecraft back onto an even keel. Only a modest pulse is required. Too vigorous a push would send the spacecraft bouncing back and forth like a ping-pong ball on concrete, wasting the finite stock of compressed gas. An exploring machine with a well-designed stabilization system will sail placidly through space with only infrequent and gentle pulses from its gas jets.

The bright and ever-present Sun is a logical spacemark for journeys in the solar system, and, although staring fixedly at the Sun doesn't sound comfortable even for an instrument, nothing prevents a simple, reliable little sensor from gazing at the shadow of a small "umbrella," which serves just as well. A Sun sensor attitude control reference system is delightfully simple in principle, sending a "restoring" signal as a shadow moves. Such a system can be made by mounting a small square shade to partially cover two pairs of identical solar cells oriented at right angles, and connected with standard bridge circuitry so that small differences in voltage outputs from matched pairs of cells produce error signals. When two matched cells are exposed to the same amounts of sunlight and shadow, they produce the same voltage output. If the shadow moves so that one receives more sunlight and the other less, the voltage difference can be used as a restoring signal to the attitude control jets. When all four sensors produce the same output, they are oriented at right angles to the Sun, thus providing two of the three axes required for stable reference.

The concept has two minor constitutional weaknesses: (1) it is necessary to preorient the spacecraft roughly in the correct direction in order for the sensors to find the Sun and become effective, and (2) if at the end of a long life the solar cells should chance to age unequally, the spacecraft could develop a list. The initial positioning of the spacecraft is made possible by sensors that determine whether the Sun is shining on the top or bottom of the spacecraft. Careful selection and quality-control processes minimize the risk of varying solar cell lifetimes.

Earth itself seemed to be a good choice for a second spacemark and was used by the Rangers and the first Mariners. It was attractive because the directional antenna needed to be aimed at Earth and the two could be aligned together. Earth proved to be less than ideal, however, for the angle it subtended varied with distance and its apparent brightness diminished greatly as the spacecraft traveled away from home, requiring a sensor of greater sensitivity. As distances grew, the sensor had trouble discriminating between Earth and the Moon, and between Earth and other planets. Earth also moved, introducing still another variable into the calculations. After Mariner 2, for which the Earth served as a workable but somewhat undependable reference, spacecraft designers switched to Canopus, a bright star in the southern hemisphere, for the second spacemark. Shining brightly in an otherwise undistinguished neighborhood, so distant it appears motionless, Canopus has been a guide star ever since for most of the interplanetary exploring machines.

This combination of sensors and systems connected to the on-off valves of the attitude control gas jets allowed stable platform orientation of the spacecraft, maintaining alignment with the Sun and Earth in inertial space so that power, thermal control, and communications needs could be satisfied. The "muscles" providing attitude control of Rangers and Mariners were cold gas nitrogen systems weighing about 4 pounds. They used a small bottle of high-pressure nitrogen and tiny jets mounted on the ends of the solar panels and the superstructure. Because of the finite quantities of gas, duty cycles had to be carefully and accurately controlled to minimize usage.

Reference to remote spacemarks must be temporarily abandoned during midcourse trajectory corrections, and it is desirable to have a temporary set of references if the spacecraft loses its lock on its distant star guides. For this purpose a three axis set of gyros is used. As already mentioned, gyros are not reliable over long intervals, being vulnerable to the accumulation of small errors caused by friction, but they are trustworthy for limited times. (A new design, the laser ring gyro using light beams is now being integrated into aircraft systems, and holds high promise of extreme accuracy for extended intervals.)

Mariner 2 was stabilized with its longitudinal axis pointed at the Sun, holding the spacecraft in both pitch and yaw directions. Roll stability was achieved with an Earth sensor mounted on the directional antenna. Pointing the long axis at the Sun provided the maximum amount of solar energy transfer to the solar panels and aided thermal control of the spacecraft by maintaining a constant Sun impingement angle, allowing the aft end of the spacecraft to point at the dark sky to radiate away excess heat. Initial Sun and Earth acquisitions were performed by internal logic circuits that derived their input from sensors and gyros.

The thermal control of the spacecraft was intended to be as passive or automatic as possible. The greatest part of the heat load came from the Sun and a lesser amount from the onboard electronics equipment, the latter also being among the most heat-sensitive components. For passive control, materials with different absorption and emission properties were used to radiatively balance the heat within the spacecraft. In addition, one of the six boxes around the hexagonal structure was fitted with louvers activated by a temperature-sensitive bimetallic element. If temperatures within the box rose too high, the louvers opened to radiate the heat to black space; when temperatures were too low, the louvers closed to keep in the heat generated by electronic components.

If Mariner's design process had stopped at this point, the craft might have been likened to a beast of burden just able to carry a small load while being led by its master. However, it would not have been able to execute a mission to Venus without some ability to plan and sequence its activities—without a kind of humanoid intelligence. The subsystem that provided these traits was the central computer and sequencer, called the CC&S. Because this was the brain center of the spacecraft, it will be mentioned often. On more advanced spacecraft, units performing similar functions have different names, but understanding the concept of the CC&S will probably enable you to communicate with spacecraft engineers.

The CC&S on Mariner supplied onboard timing, sequencing, and some computational services. Its memory contained a handful of prestored commands, and it was able to respond to a dozen specific commands sent from Earth. Since communication problems might prevent detailed orders from reaching the spacecraft, some preprogrammed intelligence was provided; after receiving the proper initiation commands, the spacecraft could then act on stored information. For example, it could acquire the Sun and Earth, going through a series of actions, after being told to. Parts of the midcourse maneuver sequence were integrated into the spacecraft memory because this was efficient and precise. Onboard sensors could determine how much the velocity had changed and could cut off the rocket after a specified increment; they could do this several minutes before engineers on Earth would have received the information from one action necessary to determine the next. Even though the journey to Venus would take more than 100 days, preprogrammed instructions for actions at encounter were also stored, so that if our command capability had been lost, the CC&S might have ordered the proper spacecraft functions. At the time, we thought that the CC&S was a marvel, little knowing how distinctly limited a brain it would seem when compared with its successors.

Basic to the CC&S was a clock that provided an accurate reference base. The clock was started during the countdown to launch, and it supplied and counted timing signals, much like today's digital watches (in fact, digital watches are an outgrowth of this space technology). Being able to count pulses, it would issue commands at set points of time throughout the mission. Packaged in a box about 6 by 6 by 10 inches, the CC&S included the highly important oscillator that provided the timing base, and it watched particularly over three critical interludes: launch (from liftoff to cruise), the midcourse correction, and encounter.

The CC&S could be given 12 different commands from Earth, although only 11 were used during the mission. It also had the capability of storing three commands that could be actuated at a precise time (not unlike sealed orders given to a ship's captain). The real time commands were specific instructions to be carried out on receipt of a coded signal, such as whether to use the directional or the omnidirectional antenna and whether to turn on or turn off the scientific instruments. To gain maximum use of the limited number of command channels, some would be used more than once by pairing one-time functions when it was possible for the repeated commands to unambiguously relate to different functions. The command for Sun acquisition, for example, was coupled with the command for unlatching the solar panels, because once the pyrotechnic squibs that did the unlatching had been fired, another command to that function would have no effect. All three stored-aboard commands dealt with the midcourse maneuver to improve the spacecraft trajectory. One told the spacecraft to roll a specified amount, one told it to pitch a specified amount, and one told it to achieve a specified velocity change.

The idea of accidental or purely random operation of the command system was horrifying, of course, and much thought went into protecting it from malice or mischance. Engineers had learned this lesson the hard way while working on an Earth satellite program in the early 1960s. In this case, a sudden rash of mysterious and erratic behavior of the orbiting spacecraft was painstakingly traced to spurious radio signals from a Midwestern taxicab. Though clearly a freak accident, the possibility of sabotage or, more likely, the inadvertent transmission of an improper or mistimed command was ever present and frightening. A complicated tamper-proof system for sending commands was devised that allowed only the correct orders from the correct people to be transmitted and acted upon. Though the system necessitated an often tedious process of reading, writing, and verifying all commands, it very likely prevented potentially ruinous mistakes.

A spacecraft on a one-way trip is useless if there is no way of sending it orders from Earth or retrieving the scientific data it collects. Like the audible or visual contacts required with a roaming hunting dog, the communications system is the only link between a distant robot explorer and its terrestrial masters. In addition to carrying orders and data between man and machine, the communications signals can be used for tracking the spacecraft, yielding startlingly precise computations of where the unseen voyager is and how fast it is going. Amazingly, Mariner's transmitter power of about 4.5 watts—less than that of a good walky-talky rig on Earth—was able to provide a communications link over a distance ranging to nearly 40 million miles.

Communications with spacecraft have usually made use of radio frequencies in the electromagnetic spectrum, although successful experiments have been conducted using lasers. For all the early lunar and planetary missions, radio frequencies in the L-band and S-band were used (about 1000 mhz and 2000 mhz, respectively). Several things affect radio transmissions: one is the distance relationship known as the inverse square law, meaning here that the strength of a given transmission signal is inversely proportional to the square of the distance from the transmitter. Another problem with transmissions to and from the surface of Earth is a serious attenuation of the signal by the ionosphere and its electrical fields. The ionosphere is a boon to shortwave transmissions on Earth because it reflects or bounces signals back toward the surface; these reflective properties tend to bounce short wavelength signals back into space if transmissions are attempted from space to Earth. To overcome this problem, we must transmit at higher frequencies that are able to penetrate this region.

A simple analog to this effect might be useful. Suppose I'm talking to you and someone places a blanket between us. This diminishes the level of the sounds reaching your ears. Depending on the type material used, this filtering might affect different frequencies more or less. This is the case for ionospheric effects: if radio frequencies are high enough, they are not attenuated so much that radio communication is inhibited. Fortunately, frequencies of about 1000 megahertz or greater are suitable for communications to and from Earth and deep space.

Another aspect of limited bit rate communications is the need to send all information in a coded shorthand language. The limitations of early spacecraft made this coding very important. Our language has 26 letters, but, compared with shorthand, it is wasteful of bits. Scientific information can be reduced for transmission and then recreated or expanded as necessary. A picture contains many bits of information and may truly be worth "a thousand words," but it is possible to compress the bits in a picture by planning. Suppose, for example, that we know the spacecraft will be taking a view showing the horizon of a planet with sky above it. If we are only interested in features of the planet and not the sky, we can program the system to send only the portion of the image containing the edge of the planet, simply discarding bits showing the discreetly differing sky. Such a technique presupposes some knowledge of the answers being sought, but it is nevertheless a useful concept.

The sophistication of the communication language depends on the type of data to be sent. One form of compressing data involves indexing transmissions against time to give different meanings to the same signals. There are many schemes that can be applied; what they all have in common is that a coded system for transmission must have a decoding system on the other end to complete the communications process.

From early in a cruise period, tracking a spacecraft leads to two calculated numbers of special interest. One predicts how far from the target planet the spacecraft will come at the moment of closest approach. Having an acceptable miss distance is vital. If the path is too close, scientific instruments can manage only a brief, blurred scan of a huge planetary disk. If the pass is too far away, as is much more likely, the instruments cannot capture all the data they were intended to collect. A desirable flyby range is assumed when the instruments are designed and sighted, and a major departure from it will impair mission results.

The second number that is examined during cruise predicts the time at which closest approach will occur. In effect, this defines the period during which the scientific instruments can reap the richest harvest. It also establishes which of the three Deep Space Network stations, located at three longitudes around the globe, will be in position during those hours to receive the explorer's signals. There may be reasons to change the time of closest approach: for example, if closest approach will occur when the spacecraft is disappearing over the horizon of one station and just rising at the next. The variable quality of equipment or of terrestrial communication links can make it desirable for a particular station to be the one to receive the spacecraft's reports during a critical time. Still another reason for adjusting the time is the angle of the Sun on the hemisphere of the planet being flown past. Pictures taken at local midnight are not very informative, and images under high-noon lighting are not ideal for showing surface relief. Any of the above factors, or a combination of them, can make it desirable to adjust the time of closest approach.

As a practical matter, neither number (time or distance of closest approach) will be perfect. Imprecision in velocity at injection, inexact assumptions about the gravitational pull exerted by the Sun and other planets, even the delicate pressure exerted by the particle streams known as the solar wind, extending over the immense length of an interplanetary trajectory, can mean that if the course remains unchanged, the spacecraft will fail to fly through the target hoop of its destination. Distance is the more important factor: there is little sense in fussing over flyby time if flyby geometry is poor. In any event, the numbers are coupled; if one is changed, the other changes. Confidence in the accuracy of the predictions grows steadily during the cruise phase. Early estimates are not altogether trustworthy, but as tracking continues and the data are integrated, it usually becomes evident that for a fully successful mission the trajectory of the spacecraft must be altered.

Assuming that the tracking accuracy provides the necessary knowledge of position and rate, a thrust vector addition can be determined. Once more a rocket becomes the means of producing a vector change in velocity. Integrating a rocket system containing combustibles under high pressure onboard a spacecraft that is carrying delicate sensors calls for careful engineering. The ideal place to put propulsion systems is at the center of gravity of the spacecraft, because as the propellant is used, the balance of the spacecraft will not be affected. Also, the thrust of the motor must be aligned such that it acts through the center of gravity, otherwise the spacecraft might spin up in space like a Chinese pinwheel on the fourth of July.

But how can this knowledge of position, velocity, and attitude be combined with the rocket thrust capability to correct the trajectory? The spacecraft is millions of miles away, moving at high speed, with only the most tenuous radio links connecting it to Earth. The solution is to execute remotely a complex *pas de deux* called the midcourse correction maneuver. The spacecraft is ordered to abandon temporarily the locks on two spacemarks that have held it stabilized in three axes, to turn according to gyro references until it is pointed in a calculated direction, and then to fire an onboard rocket of known thrust for a precise length of time. A timed rocket burn obviously depends on an Earth-based calibration of thrust under simulated conditions. We can also use inertial accelerometers to terminate thrust after the desired change in velocity has occurred. In either case, this adds a vector change to spacecraft velocity and introduces speed and angular deviations in its trajectory. The spacecraft then returns itself to cruise orientation, searching for and reacquiring lock on the two spacemarks, and continues on its corrected trajectory.

It is at the very least anxiety provoking for ground controllers to decide upon a course correction. Leaving cruise orientation is in itself a chilling step, for solar power must be abandoned, leaving nothing but finite batteries and their switching circuitry. Also abandoned for the moment is the high-gain antenna pointed at Earth, leaving only a small omnidirectional antenna certain to transmit weaker signals and capable of receiving only strong ones. Also given up for the duration of the process is the laboriously achieved thermal balance that has kept sunlit surfaces from frying and shadowed areas from freezing. The job of orienting the spacecraft with high accuracy in two planes is turned over to gyros that, sophisticated though they may be, are nevertheless intricate electromechanical devices that are heir to all the natural indispositions of the species. Then, central to the entire gamble, the rocket must work as expected, starting smoothly, developing correct thrust, and cutting off cleanly without a burp. Finally, the spacecraft must be brought back to cruise and relocked on its two spacemarks, solar power and the high-gain antenna must be brought back on line, and deviating temperatures must be eased back to normal.

To its anxious masters on Earth, the spacecraft reports by telemetry the approximate execution of all these tasks. However, telemetry cannot communicate immediately how well the tasks have been done. It takes hours, even days of tracking to enable engineers to predict with confidence the new course of the far traveler. With good fortune, skill, and patience, it will be closer to where it should be, carrying its precious cargo to the vicinity of the target planet.

The scientific instruments, sometimes thought of as the payload or passengers onboard the spacecraft, actually become integral parts of the spacecraft system. They depend on the bus for more than just transportation—they need power, thermal control, and telecommunications. As soon as they are chosen for a mission, they are integrated into the spacecraft as if they are basic components.

Although Venus is Earth's closest planetary neighbor, we knew little about it when Mariner was being planned. Men had viewed it for centuries as the brightest object (next to the Sun) in the heavens, even supposing it to be two objects because of its presence in both morning and evening. To astronomers' telescopes it was a brilliant object without much detail. Except for its crescent shape (owing to its position between Earth and the Sun during

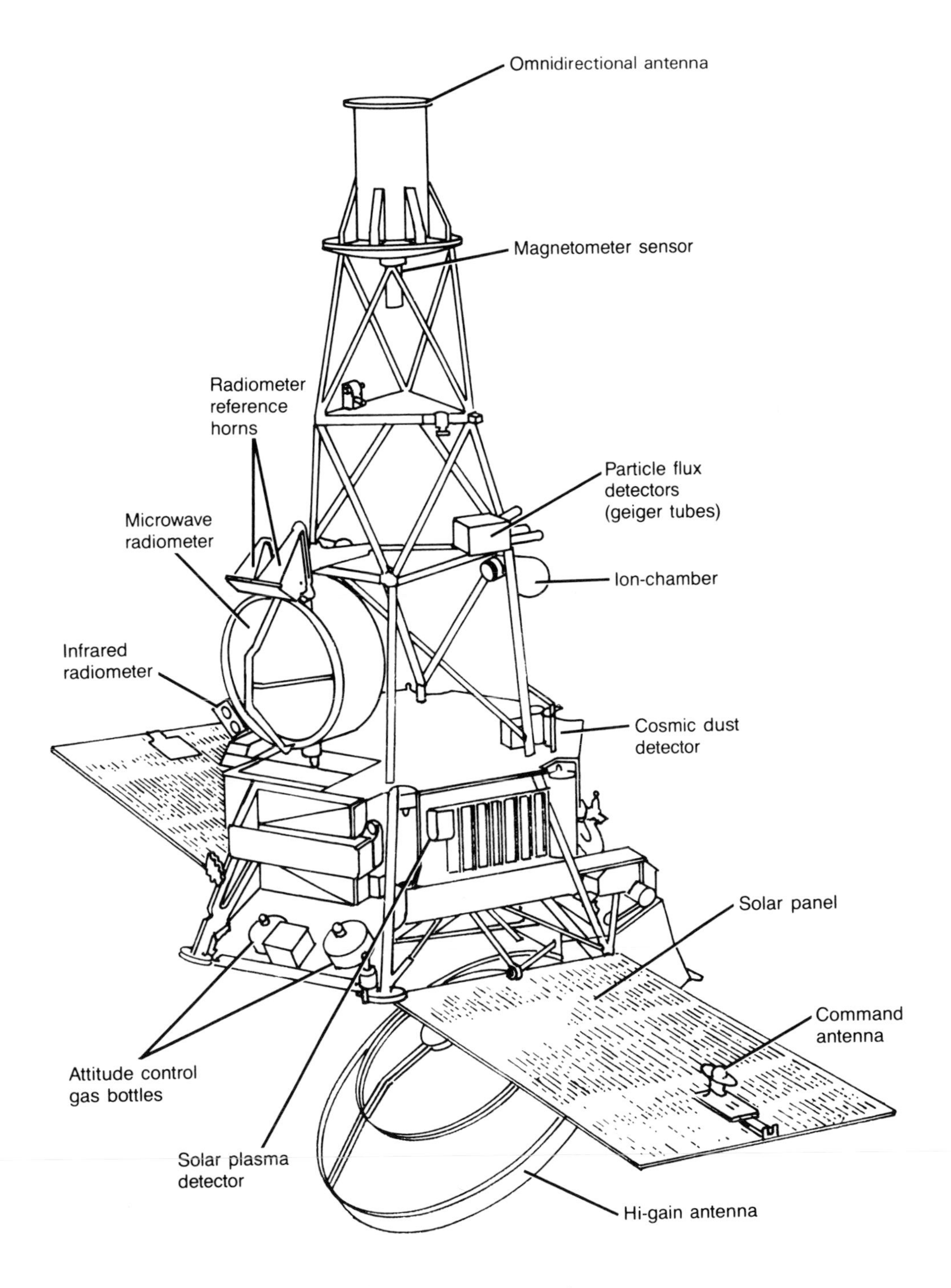

Mariner 1 spacecraft

most of its orbit), the only variations in its features were occasional changes in light and dark markings that appeared on its dense clouds (impenetrable in the small region of the electromagnetic spectrum visible to the eye), which hid the surface.

Scientific questions about the atmosphere, clouds, and temperatures of Venus were logically chosen for emphasis on the first Mariner mission. The instruments devised to address the questions were microwave and infrared radiometers: one to scan the surface at two radiation wavelengths and one to scan the clouds and give us a better idea of the cloud-top temperatures. From an engineering point of view, mechanizing the scan platform that swung the radiometers back and forth as the spacecraft flew by Venus was an entirely new development.

Four other instruments were chosen to provide information about the space environment on the way to and in the vicinity of Venus. These "double duty" instruments were a magnetometer, ion chamber-charged particle flux detectors, a cosmic dust detector, and a solar plasma spectrometer. Although the radiometers were developed specifically for the Mariner mission, the other instruments were adapted from interplanetary counterparts already in use in scientific satellites.

In abbreviated form, the elements and workings of a real spacecraft, designed to be the first official envoy from the United States, Planet Earth, to our neighbor Venus, have been described. Had it been possible to send a human, Mariner might not have been created. It was a machine that had no real consciousness; Mariner did not "know" it had two high purposes: to collect information about interplanetary space and to make scientific measurements of Venus from close by. At the time of its design and development, few of us thought about the similarities of the spacecraft to ourselves or to other living creatures.

But to those who worked on Mariner 2, conscious of the precariousness of the enterprise and the unpredictable behavior of that historic spacecraft, it was not so much a rudimentary automaton as it was a beloved partner, feverish and slightly confused at times, not entirely obedient, but always endearing.

Music of the Spheres

After Mariner 1 was destroyed, only a month remained before Venus would move out of reach. There was a lot to do to prepare for Mariner 2, and we actually had only 3 weeks, since the fourth would be consumed by the full countdown and its inevitable holds. The pad was quickly inspected to assess the damage done by the blastoff fire of Mariner 1; fortunately, damage was minor, and the necessary rewiring and other refurbishments could be accomplished in time. The causes of the failure of the first launch were soon known and the corrections found to be straightforward. Changes to the Atlas guidance antenna and the addition of a hyphen to the guidance program, once recognized, presented no significant problems. It was not deemed necessary to punish anyone on the team for the failure, for in tightly knit groups the individuals at fault usually won compassion, their consciences causing more anguish than any formal reproach. At least one company official was very apologetic, however, for he had been responsible for programming before being promoted to his management position. His colleagues ceremoniously awarded him a plaque with the missing hyphen on it.

After the traumatic failure of Mariner 1, those at Cape Canaveral would have liked a few months to wind down, regroup, and carefully prepare for the second launch. However, we had to reckon with an absolutely firm deadline—set not by an overzealous program manager but by the strict geometry of our solar system. There were 4 weeks in which to complete our task and no more; a delay beyond that point amounted to certain failure. This time restriction, while especially frustrating in the summer of 1962, is a fundamental problem for any planetary mission.

Although diagrams often show the planets in neat, circular orbits around the Sun, the actual geometry of the solar system is tremendously more elaborate. The Earth and its sister planets revolve about the Sun in unique ellipses, each moving in a separate plane and with varying, precisely changing velocities. Considering the multitude of factors involved, the orbital relationships among the planets are beautiful in their geometry but bewilderingly complex. Thus, in sending a spacecraft from one planet to another, proper orientation of the launch vehicle is not enough to ensure an acceptable flyby distance. We cannot simply point a rocket toward Venus, fire it, and expect the payload to neatly sideswipe its planetary target. Once in flight, a rocket-launched spacecraft itself becomes a planet, obeying the same curvilinear laws of orbital behavior. For a successful interplanetary launch, proper timing is every bit as important as correct orientation.

The best opportunities for launching a spacecraft from Earth to Venus occur once every 19 months. During these short periods (about a month or so in duration), the energy required for a successful launch is at a minimum. If a voyage were attempted at any other time, the amount of energy needed would be prohibitive. The reason is that while Earth and Venus both circle the Sun in near-circular orbits, the spacecraft must leave Earth and travel in its own elliptical orbit about the Sun, arriving at the orbit of Venus at a time when Venus is also there. Since Venus moves faster around the Sun than does Earth, it would move from behind to ahead of Earth during the total interplanetary flight time of Mariner 2.

In addition to the approximately month-long period when Venus is accessible, there is a daily window amounting to less than an hour. This additional complication is caused by Earth's rotation about its own axis. We were able to extend the window slightly by varying the timing delay in the parking orbit before the Agena's second burn, but this nevertheless imposed a tight restriction on the last part of the countdown.

Mariner 2's countdown began on August 25 at launch time minus 205 minutes. The spacecraft had by then accumulated a total test time of 690 hours—better than 4 full weeks of operation in which its parts had been given a chance to fail and had demonstrated their likelihood of lasting through the long trip in space. Soon, however, the count had to be scrubbed because of an indication of stray voltage in the Agena destruct circuit. This was corrected and the count restarted on August 26. In all, there were four unscheduled holds, and the launch was delayed a total of 98 minutes. There were several tense interludes in those last hours, particularly after the Agena had been loaded with propellants. If the daily window had been missed, the vehicle would have had to be detanked and elaborately purged of its volatile chemicals. The Atlas battery also proved worrisome as launch time neared, for it had been replaced once and its replacement was down to a life expectancy of 3 minutes at the moment of liftoff.

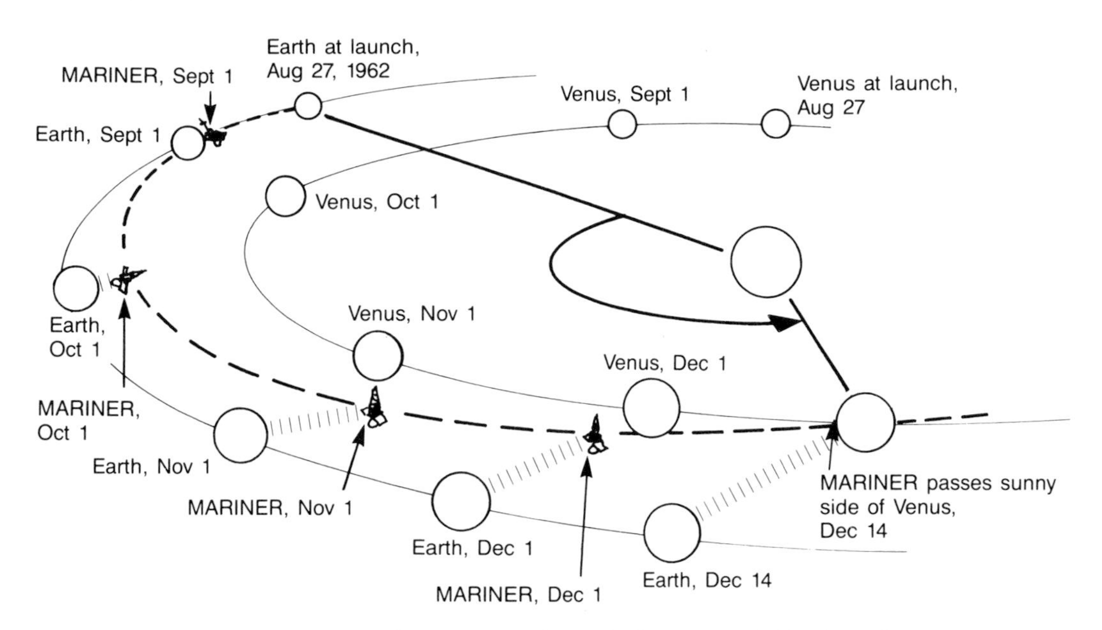

Relative orbital positions of Earth, Venus, and Mariner 2

Looking back on those Cape launches in the early 1960s, I remember human aspects fully as much as mechanical mischances. A nagging feeling of helplessness plagued me, a former "hands-on" engineer, now a Headquarters official, because of my total dependence on others, including many persons I had never met. My only recourse was to do everything possible to ensure team spirit and singlemindedness of purpose.

One idea that became a tradition involved a visit to the blockhouse in a 30-minute, built-in hold, about an hour before launch. During the Mariner 2 countdown, I accompanied the project manager and one or two other officials from the spacecraft control station in Hanger AE to show our "colors" and to wish the launch crew well during those final critical minutes before launch. In addition to helping ease the tension of the count, I knew deep inside that sharing moments when things were going well might make it easier to work together should things go wrong. Those early blockhouse visits are remembered as warm and friendly interludes during times of considerable anxiety—a chance to share "Hellos" with the Launch Conductor, Orion Reed, and other respected friends who played vital roles in the operation. Whether or not these visits had a good effect on team spirit or were otherwise helpful is uncertain, but they at least gave us something to do during the tense time of waiting that was, for me, the hardest part of the job.

I think the idea of visiting the blockhouse during the T minus 1 hour hold came from Jack James, Mariner Project Manager. Jack was extremely sensitive to the importance of each team member in the launch operation and genuinely cared about peoples' feelings. These qualities made him an excellent choice to lead operations at the Cape where Air Force, NASA, JPL, and contractor personnel were trying to do a most difficult task as a relatively unrehearsed team. The concept of a project manager and field center project management assignments was new and being used for the first time. Jack's powers were largely dependent on his ability to develop respect and on his skills in persuading people to do their tasks in a coordinated manner. The project manager was answerable to NASA for the entire operation; however, he had "hiring and firing" control over only a few staff members—the thousands of others who worked on the project did so through a complex of indirect assignments.

Jack had grown up in Texas. He confessed to a certain amount of bumming around and doing odd jobs after finishing high school; when he finally decided to enter college at Southern Methodist University, he chose electrical engineering for "default" reasons. When I asked him to explain, he said he knew that civil engineers built roads and bridges and that he didn't want to do that, mechanical engineers dealt with big machinery and he didn't want to do that either, but since he didn't know anything about what electrical engineers did, he decided it might be fun to learn and chose to enter the field. It's clear now that he made an excellent choice; solid-state physics began opening the field of electronics with semiconductor developments about the time he was in school and within a few years revolutionized our lives.

Jack's early assignments at JPL involved missiles being developed for the Army. His experiences with missile launchings were good preparation, but compared with Mariner operations involving literally thousands of people, missile launchings were games. Coupled with the usual "people" problems were the unusual schedule demands dictated by planetary orbit constraints. Getting diverse groups to work together was tough enough; getting them to mesh their efforts to meet an unyielding deadline was far more challenging. No allowances could be made for one element of the team to slip its schedule. This extra dimension, added to an effort that had never been approximated before, called for a high level of insight to plan and direct technical and human activities.

Jack had an unusual blend of experience, horse sense, and humanistic qualities. We were extremely lucky to have him around, for it was apparent by the success of the early Mariners that he was a singularly gifted project manager, one of the best we ever had. He became, in a shy, winning way, almost a national resource: technically astute, uncanny with schedules, and masterful with people.

We did not always agree on everything, but my respect for Jack's abilities made it possible for me to compromise without distress. One such compromise sticks in my mind, when we disagreed on a "judgment call." It seems relatively unimportant now, but I was concerned at the time. Our disagreement arose over Jack's plan to have one of the shiny aluminum covers on a Mariner Mars spacecraft compartment embossed with the seal of the United States. He had a cover made up with the seal so we could see what we were talking about, but since it was fastened by only a few screws, the final decision on whether to use it could be made at the last minute. His view was understandable; we were competing with the Russians in the race to the planets, and Americans could be proud that our "trademark" would be exhibited for current and future generations to see. My concern was that we might be accused of exhibitionism, something distasteful to me, for I was deadly serious about doing the mission for other reasons. The Russians had bragged about landing a pendant on the Moon, and I wanted no part in that disgusting game.

Since the shiny aluminum surface was important for proper thermal control, I questioned whether the embossing, however light, might negatively influence the thermal properties. Jack agreed that tests would be made with the panel in place. Since my greatest concern was that critics would misinterpret this symbol as a lack of seriousness on our part, I further insisted on a low-profile, no-publicity approach for the addition. The panel with the seal was installed, the tests were made, and even after the successful flight there was very little publicity about the seal, and none at all negative.

On our return to Hanger AE after the blockhouse visit, the final 60 minutes of the countdown began. These were the tensest moments of all, as the final checkouts began in earnest. During this period emphasis shifted from what had been predominately launch vehicle activities to include readiness reports from down-range tracking stations, from Deep Space Network stations, from weather stations, spacecraft operations, and all the elements required directly after launch. At this point, anyone caught bluffing would have been a traitor, yet any group not ready to go would have provided a reason to scrub the launch. The readiness reports at T minus 5 minutes were critical; when all of these were "Go" we were fairly certain that we had a commitment to ignite the rockets and accept our fate.

On this day everything came up "Go." Liftoff appeared normal, and the distinctly frightening aspects of this launch did not occur until a few seconds before booster engine cutoff. Suddenly, control of one of the two vernier engines on the Atlas was lost for an undetermined reason, and it moved to its maximum negative mechanical stop. The main booster engines compensated and were able to hold proper roll control until they were cut off and jettisoned. Then the vehicle began to roll, slowly at first and then faster. After about 60 seconds it was turning at a rate of nearly one full turn per second in an uncontrolled, unprogrammed prelude to disaster. About 10 seconds later this aberrant behavior ceased, and the launch vehicle stopped spinning, coming to rest only a degree and a half off its proper roll position. This random restoration so close to nominal has never been plausibly explained, although Wernher von Braun may have been right when he suggested that the success during this part of the mission could only be explained by a substitution of "divine guidance" for the malfunctioning Atlas guidance system.

The launch vehicle wasn't finished with its eccentricities, however. Somewhat high at booster engine cutoff, the rocket was pitched about 10° upward, and there was a slight error in azimuth. During the period of uncontrolled roll, the poor crazy Atlas had been able to respond effectively to its guidance commands—something no one would have thought possible. Separation of the Atlas and Agena occurred successfully, although the pitched-up attitude meant that the ejected shroud came perilously close to striking the Agena. In an effort to correct its attitude, Agena pitched down 2° at the start of its first burn, which prevented the horizon sensors from sensing and correcting the error until 15 seconds had passed. Further complicating matters, the excess height of the Atlas had caused the Agena start signal to be sent 8 seconds early. Then, at last, something worked precisely right. The velocity meter aboard the Agena, one of several ways to cut off the engine, sensed achievement of the proper velocity and terminated the first burn, leaving the Agena and its spacecraft in a good 115-mile parking orbit. All who knew what was going on (most of us did not) were breathless after this series of narrowly averted catastrophes.

The second Agena burn began on time and was later cut off crisply by the same perfectly operating velocity sensor. Gentle springs separated the spacecraft from the Agena, which then reversed its attitude and discharged its remaining propellant, using this residual impulse in the opposite direction to change the course of the stage and eliminate any chance that the stage itself could impact Venus. On its own at last, Mariner 2 began the long voyage to Venus.

The combination of optical and electronic tracking devices gave information suggesting that the Mariner 2 boost phase was generally satisfactory, even though the Atlas had apparently rolled 36 times during its operation. There were some anomalies in the coverage, but information was collected so that engineers could decipher what had happened at a later time. The most critical events of engine ignition and cutoff, and separations of the shroud and spacecraft, appeared to be satisfactory. When reports of all these near-Earth orbit events were complete, there were great sighs of relief from members of the project team. With all the uncertainties of the launch accomplished, it now seemed that we had a chance of success. Perhaps this feeling of relief was connected with the helplessness I felt during the launch phase, when so many strangers were involved and I had no insight as to how well they would do their jobs. Now the launch was over, and responsibility clearly rested with members of the JPL Project Team, with whom I had worked more closely.

Injection into interplanetary trajectory occurred about 26 minutes after liftoff; it was then about 5 more minutes until the Deep Space Instrumentation Facility (DSIF), using its big dishes, obtained contact with the spacecraft. From this time on, virtually continuous contact was maintained with the spacecraft until the end of the mission over 4 months later.

Critical spacecraft operations now began. Approximately 18 minutes after injection, the solar panels were extended. Full extension occurred within 5 minutes after the CC&S sent its command; this was considered nominal. The initial telemetry data indicated that the Sun acquisition sequence was normal and was completed approximately 2½ minutes after command from the CC&S. The high-gain directional antenna was extended to its preset Earth acquisition angle of 72°. The solar power output of 195 watts was slightly above the predicted amount, providing an excess of 43 watts over the spacecraft requirements for this period of flight near Earth. Although temperatures were somewhat higher than expected, they slowly decreased; 6 hours later the temperature over the entire structure had stabilized at about 84° F.

With all subsystems apparently performing normally, with the battery fully charged, and with solar panels providing adequate power, a decision was made on August 29 to turn on cruise science experiments. These had been deliberately left off for about 4 days after launch to ensure that all the atmosphere in the compartments had escaped (this is referred to as outgassing), so that electrical arcing would be unlikely. Leaving the instruments off also allowed the batteries to charge and all existing power to be applied to the engineering information and commands. When finally turned on, the cruise science instruments appeared to be operating normally in all respects. Though this was promising news, only 25 percent of the total components were exercised during cruise, and there was no assurance that the other scientific instruments, those critical ones devoted solely to the planetary encounter, would ultimately function as well.

Five days after launch, temperatures had stabilized within tolerance limits, tracking had been continuously maintained with two-way lock, telemetry data were good, and all subsystems appeared to be operating as intended. For the first time the project team, now regrouping at JPL, began to feel the effects of the year of concentrated effort plus the satisfaction of successfully initiating the mission. Back in Washington I also felt good about the start toward Venus, but after so many bad experiences in the past, I could not let myself relax and enjoy the momentary success. Maybe things would continue to work, but it was a long trip to Venus and many chancy things had to happen properly before we would have a successful mission.

About 3 days later, the Earth-acquisition sequence was initiated by the CC&S. The Earth sensor and the gyros were turned off, cruise science was turned on, and roll search was initiated. At that time the spacecraft was rolling at a rate of about 720° (two revolutions) per hour. Indications were that the directional antenna and Earth sensor were pointed 72° below the Earth-spacecraft plane, apparently because of a switch from the omnidirectional antenna to the directional antenna, and telemetry data were lost until Earth lock was reestablished 29 minutes later. At that time, acquisition data indicated an Earth brightness intensity measurement significantly lower than expected and comparable to that which would have resulted if the Earth sensor had been viewing the Moon. There was a possibility that the Moon had been acquired, implying a malfunction in the antenna hinge servo. As a result, execution of the midcourse maneuver sequence was postponed until the following day, when it could be determined that the antenna actuator had performed properly and that the directional antenna was pointing at Earth, even though the signal seemed weak.

Tracking data indicated that the launch vehicle had provided a near-nominal orbit, so that there was plenty of capability in the Mariner 2 midcourse motor to perform its correction. The midcourse maneuver was initiated on September 4 and completed early on the morning of September 5, when the spacecraft was about 1½ million miles from Earth.

The maneuver sequence required five commands. Two were real-time commands and three were stored. Commands sent directly from the ground ordered the changeover from the high-gain antenna to the omnidirectional antenna so that data could be received during the maneuver when the attitude of the spacecraft was not favorable for pointing the high-gain antenna. In addition, the high-gain antenna had to be moved out of the way so that it would not be affected by the rocket firing. The Earth sensor used for pointing the antenna was turned off so that the entire operation of the high-gain antenna was disabled intentionally by command. The stored commands necessary for orienting the spacecraft and for firing the midcourse motor were determined from trajectory calculations. The commands, sent to the spacecraft for storage in the CC&S until the proper clock time, contained roll and pitch turn durations and polarities, plus the velocity increment to be used for cutoff of the midcourse motor.

The spacecraft performed its maneuvers and provided the general telemetry data. All maneuvers, plus the burning of the motor, appeared to be normal. The entire midcourse correction took approximately 34 minutes. Telemetry data were lost for approximately 11 minutes because the spacecraft moved into an attitude where there was a partial null in the propagation pattern of the omnidirectional antenna. This was simply a feature of the particular orientation required for the midcourse motor burn and not a cause for concern. Initial telemetry data received after the midcourse maneuver indicated that all subsystems were still operating normally. In the Sun reacquisition sequence initiated by the CC&S at the nominal time following the maneuver, the autopilot used during the course correction was turned off and the directional antenna moved to the reacquisition position of 70°. The Earth reacquisition sequence was also initiated by the CC&S at the nominal time following the maneuver and again required approximately 30 minutes, with the spacecraft rolling almost one complete revolution before Earth lock was established. The transmitter was switched to the high-gain antenna at the start of the sequence, just as in the initial Earth acquisition sequence, causing severe fading and a loss of signal for approximately 6 minutes while the high-gain antenna pointed in other directions than toward Earth. The spacecraft returned to the normal cruise mode of operation with all readings similar to those obtained prior to the maneuver, with the exception of the propulsion subsystem, which had expended itself in accord with its charter.

During the period between Mariner 2's launch and its encounter with Venus, I was extremely busy and glad of it. Notes I kept show that the week following the midcourse correction was filled with activities, many about to become major challenges. Hughes Aircraft officials came to see me to express concern over Surveyor and to urge more direct involvement by Headquarters; they felt the need for three-way meetings between Headquarters, JPL, and themselves, which we finally initiated after real trouble developed. I attended meetings that addressed the Centaur launch vehicle development problems and impacts on Surveyor; others with the Space Science Steering Committee were aimed at intelligently reducing the instrument complement to be carried by Surveyor because of reduced Centaur performance. A power struggle was developing between Goddard Space Flight Center in Greenbelt, Maryland, and JPL over which center would do the planned Mars '64 missions. I was seriously interested in having a healthy competition between two capable groups, but I did not like the way the battle lines were being drawn. Goddard wanted the entire spacecraft to be a spin-stabilized capsule system, and JPL preferred a three-axis stabilized bus and capsule system. I would have preferred a cooperative approach using the talents of both centers, but that was not in the cards. This was the beginning of a long and bitter struggle between the two centers over planetary assignments, with me in the middle.

It was also during that week that I initiated the first serious discussions on how to get a lunar orbiter program going. A Surveyor orbiter concept had bogged down, and I began looking at simpler spinner spacecraft concepts for achieving this desperately needed mission. On top of these concerns over future activities, Homer Newell, Associate Administrator for Space Science and Applications, was urging me to: (1) produce more creative coverage for public information on Mariner's progress, (2) initiate preparations for congressional budget hearings, and (3) draft "white papers" on the rationale for future programs and on management aspects of our Headquarters interfaces with the field centers.

The first malfunction onboard Mariner occurred in the midcourse propulsion system following the completion of the correction maneuver. After the spacecraft motor had been commanded to shut off, the pressure reading in its propellant tank continued to rise. It was presumed that the normally open nitrogen shutoff valve did not close fully as the motor was shut off, letting nitrogen gas leak slowly into the propellant tank. A quick calculation showed that the equilibrium pressure, when reached, would be well below the burst pressure of the propellant tank and associated components. Accordingly, no further complications were expected or observed, since the high-pressure nitrogen was simply leaking into another tank and not escaping from the spacecraft. The leak had no effect on the attitude control or any other function of the spacecraft, and since no further rocket firing was planned, no corrective action was taken.

Post-midcourse trajectory computations indicated that Mariner 2 would miss Venus by approximately 25 000 miles and that the flight time for the entire trip would be about 109½ days. A comparison of the desired and achieved encounter parameters indicated that the midcourse maneuver was accomplished with near-nominal performance. There were a number of possible explanations for a slightly out-of-tolerance correction, but telemetry data could provide no clear clues that would isolate the cause. While we would have preferred being closer—more like the planned 18 000 miles—it was believed that 25 000 miles was well within predetermined values for instrument design. As the spacecraft approached the planet and more tracking data were available, trajectory predictions showed that the actual miss distance would be 21 645 miles.

About 3 days after the midcourse maneuver, telemetry information showed that the autopilot gyros had automatically turned on and that the cruise science experiments had automatically turned off, possibly because of an Earth sensor malfunction or an impact with an unidentified object which temporarily caused the spacecraft to lose Sun lock. All attitude sensors were back to normal before the telemetry measurements could be sampled to determine whether an axis had lost lock. A similar occurrence was experienced 3 weeks later when the gyros were again turned on automatically and the cruise science experiments were automatically turned off. Here again, all sensors were back to normal before it could be determined which axes had lost lock. By this date, the Earth sensor brightness indication had become essentially zero. The significant difference between the two events was that in the second case, telemetry data indicated that the Earth brightness measurement had increased to the nominal value for that location along the trajectory. This problem remained a mystery and was somewhat worrisome because of the possible implication that the attitude control system sensors were marginally effective in maintaining the desired attitude references.

On October 31, there was an indication that Mariner's power production had decreased, a malfunction later diagnosed as a partial short circuit in a solar panel. As a precaution against the possibility of the spacecraft rapidly sapping its battery stores, a real-time command was transmitted from the Goldstone station in California, turning off the cruise science experiments and thereby reducing power consumption. The lower power condition existed for some 8 days, when suddenly the telemetry data again indicated that the panel was operating normally. After this was confirmed, another command was transmitted from Goldstone to reactivate the cruise science experiments. The science telemetry data remained essentially the same as before the experiments had been turned off; however, engineering telemetry data indicated that most temperatures had increased shortly after the science experiments were reactivated, probably due to the increased power requirements. A recurrence of the panel short was experienced on November 15; by this time, however, the spacecraft had proceeded nearer the Sun and the power supplied by the one operative panel was enough to meet the spacecraft's needs; thus, the cruise science experiments were permitted to remain active. Along with this anomaly, the magnetometer experienced a high offset, probably caused by a current redistribution when the power failure occurred. This made readings more difficult to interpret, but the recorded data indicated reasonably steady magnetic fields.

The radiometer calibration performed during the cruise phase indicated that the instrument would malfunction when activated for the flyby of Venus. It was considered possible that when the cruise science mode was changed to the encounter sequence the radiometer would remain in a permanent slow-scan mode, and no high-speed scan or automatic scan reversal would occur. In addition, the telemetry data indicated that only one of the two microwave radiometer channels would have the desired sensitivity. It turned out, however, that both the microwave radiometer and the infrared radiometer channels had acceptable sensitivities at encounter and that one scan rate change occurred, allowing three successful scans of the planet.

Another worrisome problem in the scientific instrumentation was soon detected. On November 27, the calibration data for the cosmic dust experiment indicated that either the instrument sensitivity or the amplitude of the calibration pulse had decreased by 10 percent. By December 14, a further decrease by a factor of 10 had occurred. These figures suggested that the instrument's operation would be severely impaired at the time of flyby.

Mariner's technical troubles were not limited to malfunctions onboard the spacecraft. On one occasion, a commercial power failure at one of the tracking sites caused the loss of 1½ hours of data. In mid-November, an occasional out-of-sync condition in the telemetry data was determined to be the fault of a telemetry demodulator at the tracking stations and not of the spacecraft's instrumentation. No real-time telemetry was transmitted from Goldstone and Johannesburg during the November 26 view period. The information was not lost, however, since all data were recorded on magnetic tape at each station and could later be sent to the Space Flight Operations Facility for full processing.

Except for problems of this nature, the DSIF stations covered the Mariner 2 operations continuously and successfully. In taking two-way Doppler data for orbit determination, one Goldstone antenna transmitted to the spacecraft and the other received signals from the spacecraft. On one occasion, the spacecraft antenna reference hinge angle changed slightly, an event which should have occurred only at cyclic update times. This phenomenon had appeared several times during preflight system tests and was not considered serious. With the exception of this anomaly and the Earth sensor anomalies noted earlier, the attitude control system performed without fault through the mission.

In mid-November, spacecraft temperatures became a cause for concern as they began to exceed predicted values. On November 16, the temperature of the lower thermal shield reached its telemetry limit and pegged—this corresponded roughly to 95° F. Seven of the eighteen temperature measurements were pegged for the encounter phase, and the actual temperatures had to be estimated by extrapolation. It seems that spacecraft, like people, suffer at times from extremely high temperatures; there was considerable concern that electronic components would be adversely affected by this condition. On December 9, a failure in the data encoder circuitry disabled four telemetry measurements: antenna hinge angle, propellant tank pressure, midcourse motor pressure, and attitude control nitrogen pressure. Though the loss of these particular measurements did not affect the outcome of the mission, failures of this sort were deeply troubling to the project team. Not knowing exactly what caused these malfunctions, we worried that more critical systems might suddenly (and inexplicably) begin to deteriorate.

The CC&S was designed to perform various functions, one of which was to provide the attitude control subsystem with a timing or cyclic update to change the Earth-pointing antenna reference hinge angle. Each cyclic update pulse was indicated by telemetry. Until December 12, the pulses occurred with predictable regularity. On that day, however, only 2 days before the encounter phase, the CC&S failed to issue the 155th or subsequent cyclic pulse. As a result of this malfunction, the spacecraft was switched on December 14 to the encounter mode of operation by a prearranged backup command transmitted from Goldstone. Just prior to this transmission, seven spacecraft temperature sensors had reached their upper limits. The Earth sensor brightness data number had dropped, and approximately 149 watts of power were being consumed by the spacecraft. About this much power was available from the one good solar panel, and a small excess of about 16 watts was actually being dissipated. All science experiments were operating, and coverage by the DSIF remained continuous and appeared normal. Signals were clear, and data quality was good. Of course, there was considerable concern over the fact that several minor failures had occurred in telemetry measurements, the failure of the CC&S update of the antenna, and the associated possibility that the scientific scan platform might not operate as designed. With the spacecraft running a high fever, the preencounter hours were extremely tense.

We will probably never know for certain what went wrong inside the CC&S, although higher than expected temperatures surely played a part. It was suspected that a single component was the culprit. Within the region where the failure was isolated there were 160 resistors, 51 transistors, 50 cores, 40 diodes, 25 glass capacitors, and 4 tantalum capacitors. Any single one of these could have been the cause.

The operation of all science experiments during encounter was essentially as planned, except for the sensitivity decrease in the cosmic dust experiment. The encounter mode lasted approximately 7 hours, being terminated by a ground command from Goldstone at 20:40:00 GMT on December 14, 1962. Engineering telemetry data transmitted after the encounter phase indicated that all systems appeared to be performing essentially as before. However, temperatures continued to rise and were not expected to decrease as the spacecraft was approaching the Sun, scheduled to arrive at its perihelion or closest approach on December 28.

As a result of the CC&S malfunction for orienting the Earth-pointing antenna, the antenna reference hinge angle had not been updated since December 12. Since the change of angle was slight during the period of encounter, no correction had been necessary. Afterwards, however, it became apparent that some adjustment was needed to ensure continued communications between the spacecraft and Earth. Two series of commands were transmitted from Goldstone, on December 15 and December 20, updating the reference hinge angle. Five of these commands were accepted and acknowledged by the spacecraft, and an effective reference angle change of 8° occurred as desired.

On December 17, after an extremely busy 3-month period, the continuous coverage of the DSIF was reduced to approximately 10 hours per day to provide relief to overworked personnel. As expected, perihelion occurred on December 28. On this date an attempt was again made to command the reference hinge angle to change, but Goldstone was unable to lock up the command loop, indicating that command thresholds had been passed. On December 30, a reference frequency circuit failure in the CC&S countdown chain resulted in a temporary loss in telemetry; however, radio frequency lock, that is, the closed-loop coupling of the spacecraft transmitter and the ground station receiver, was maintained. When the telemetry signal was again acquired 1½ hours later, the telemetry bit rate had dropped from the nominal 8.33 bits per second to approximately 7.59 bits per second. Simultaneously, internal temperature readings increased due to the inefficiencies of the power system at lower frequencies.

The spacecraft was tracked for the last time at 07:00:00 GMT on January 3, 1963, by the Johannesburg station. During this pass, about 30 minutes of real-time telemetry data were received. Although the demodulator went out of lock and remained out during the later part of the tracking period, good tracking occurred for most of the interval. Examination of the recorded data showed that the spacecraft was still performing normally, with a power consumption of 151 watts and available power of 163 watts from the single operating solar panel.

In the final review of the orbits, the spacecraft was last heard from when it was 53.8 million miles from Earth and had passed Venus by about 5.6 million miles. It was traveling at 13.7 miles per second with respect to Earth and disappeared at this time, never to be heard from again. Further searches for the spacecraft at later periods were unsuccessful. On January 8, 1963, the Goldstone antenna was positioned according to the projected trajectory data, and a frequency search was conducted during the calculated view period, with negative results. A similar attempt in August 1963 was also unsuccessful.

Thus ended the saga of Mariner 2—a robot, designed and directed by men, given a mission to extend the search for knowledge beyond the limited reach of *Homo sapiens.* Though it accomplished a voyage that was clearly "superhuman," Mariner was a simple exploring machine, with only a very modest capability to perform on its own. A total of 11 real-time commands and a spare were possible, along with a stored set of 3 onboard commands which could be modified. Other functions, such as updating antenna position and adjusting thermal control louvers, were provided, but in every sense it was a simple robot with the capability for only a small amount of human interaction.

From the meager information returned by telemetry, we know that Mariner 2 endured significant stress, but how many meteorite impacts it received and why it developed an ultimately fatal fever will forever remain a mystery. Perhaps in its passage from Earth to Venus and its transfer from orbit to orbit, it had other experiences which we will better understand when man repeats the voyage in person, with his own sensors and the additional capabilities that will exist at the time.

However modestly equipped to observe the environment and features of Venus, Mariner 2 did provide to those on Earth a firsthand, close-up impression of Earth's nearest neighbor—a brilliant object long revered as the star of the morning and evening. Indelibly imprinted in my memory is the beautiful sound of the data stream returning from the encounter science experiments during flyby. The radio telemetry signals were transmitted at L-band frequencies of about 940 megahertz and reproduced as whole tones well within the audible range. These tones were broadcast throughout the operations facility and relayed to NASA Headquarters for all to hear during an encounter press briefing. The pure tones at the low bit rate of 8⅓ bits per second produced heavenly angelic sounds, truly music of the spheres. Words could not describe my feelings as the successful return of data from Venus at last provided evidence of a successful mission.

During the time the Mariner 2 spacecraft was on its way, a fifth Ranger mission had failed and the entire Ranger program had been interrupted for review and possible cancellation. Thus the Mariner 2 flyby was a victory of far greater significance than its single-purpose objectives, for it gave some confidence to the process and to the team efforts involved in developing the capability for such exploration. Because of the "Review Board" environment and the large amount of work that resulted from the Ranger failure, there was little time to revel in the Mariner success; however, I remember that Christmas of 1962 as one of the better ones during my early years with NASA.

Jack James' penchant for patriotic display came to light as Mariner 2 was well on its way to Venus, when he disclosed that he had personally placed a small American flag between some layers of thermal material on top of the spacecraft. Had I known about this when it occurred, I would have reacted as I did when Jack later had a seal added to the Mariner 4 compartment cover.

Some day future Americans may recover Mariner 2 and rejoice in exposing its national symbol, proving that Jack was right in doing what I considered to be sensitive at the time. As things turned out, I am proud that our flag and great seal are out there in orbit about the Sun along with the planets.

Looking back on the entire experience, my warmest feeling comes from the association with the crew that produced the Mariner mission. Starting with the handful of us involved in directing the program at NASA Headquarters, the project team numbered about 250 at JPL, spreading to 34 subcontractors and over 1000 suppliers of parts for the Mariner systems. Altogether the project involved an estimated 2360 man-years of effort and cost a total of $47 million. At the time, so much effort and so many dollars expended in a year seemed large. Today, we have learned that such a price is relatively small for the results returned. Not only did the thousands of people who participated in this "once-in-history experience" gain from it, but Americans and all mankind received a boost in spirit from the adventure.

The first successful Mariner mission will surely become legend, remembered as a triumph for creative man. As someone aptly put it, "There will be other missions to Venus, but there will never be another *first* mission to Venus."

The Basis For It All

The sailing ship evolved during the early years of the Renaissance. At first men were able to navigate rivers and streams with canoes and small craft; later they were able to explore the oceans. Their ships had to be large enough to carry crews for manning sails and covering watches, and supplies to sustain the men during voyages that lasted several months. Staying afloat for long periods of time was not enough; these ships had to be rugged to withstand the gale winds that were sure to come with long exposures at sea and capable of staying underway around the clock. The propulsion technologies for these craft evolved from men at oars, to sails that allowed only downwind motion, and then to sails that allowed tacking in chosen directions. This evolution took place over many hundreds of years, with the greatest advances occurring during the periods when men were motivated to explore Earth.

The ability to withstand the rigors of the seas and to master the winds, while necessarily first, would not have allowed the systematic exploration of the oceans and distant continents had it not been for the development of the compass. The compass, a technological discovery that provided a known direction any place on the seas, was not only a help in its direct navigational capability but surely gave the sailors greater confidence. With this device they not only knew which direction they were going, but could always find their way home.

The invention of the clock made it possible for navigation to become more than just determining direction; position could be known as well. With the combination of the chronometer and the compass it became possible for sailors to determine directions, positions, and rates of speed with the precision necessary to navigate predictably across the oceans.

Even given the ships, the life support for the sailors, the propulsion of the winds, the navigational tools, and other necessary technologies, exploration would not have occurred had additional factors not been at work. A major motive for exploration was the thirst for knowledge, for riches, for discovering what was there. In addition to these basic human drives, competition with other nations and the prestige of those who were able to sail to the far corners of the Earth and return home with evidence of new conquests and discoveries stimulated this activity. Many of the sailors who went on such voyages did not go voluntarily; zealous leaders were sometimes able to acquire prisoners provided by heads of state and were willing to take such men and launch into unknown regions with the expectation that crews could be whipped into shape while performing the necessary services.

Except for the need to commandeer personnel, space exploration in the 1960s and 1970s required exactly the same basic ingredients. Like the sailors who gazed longingly across the oceans for centuries before the ship technologies evolved, men studied the heavens and dreamed of visiting the Moon and the planets before launch vehicles and spacecraft were feasible. Had the technologies been available to them, they would surely have tried to do the things we have so recently accomplished. Our serious activities in space had gotten underway before he was elected president, but John F. Kennedy was to forever remind us of the similarity between the exploration of Earth and space when he used the haunting words in his inaugural address, "We set sail on this new ocean. . . ." Our blessing is that our generation was privileged to experience that goal.

Invented in China at least as early as the 1300s, rockets have been generally understood for centuries. For most Americans, however, the bombing of London with V-2s brought the shocking realization that rockets could do things we had not believed possible. As an aeronautical engineering student at the time of the first V-2 bombardment of Britain, I was absolutely amazed at the capability of any vehicle to go 3500 miles an hour or as high as 100 miles above Earth. After all, we had been taught that "compressibility effects" at the speed of sound were deterrents to high-speed flight in the atmosphere, and that aircraft would not likely ever fly more than 600 mph because of the so-called "sound barrier" and the heating involved. Yet suddenly, here were vehicles traveling many times that speed and at altitudes far greater than the atmosphere that limited flight from an aeronautical standpoint.

It was not long after the V-2 reports that I learned of the efforts of Robert H. Goddard in the 1920s and 1930s that led to the development of the liquid rocket and the interest of the Germans in the application of this technology. However, it was a while longer before I learned how amazingly simple the whole idea of a rocket was, for there had been such a preoccupation with aeronautics between World Wars I and II that few engineers and none of my professors had even thought much about them. This was the first of several instances in which I was surprised to discover that existing technologies were simply overlooked for a period of time before engineers began to make good use of them.

A related case is the gas turbine or turbojet. When I was forced to take a thermodynamics course concerned mostly with steam turbines as part of my engineering curriculum, I was distressed because I thought it was a waste of time for an aeronautical engineer to study ground-based power plants. Within 5 years turbojet engines were revolutionizing aeronautics, raising the obvious question, "Why in the world hadn't the steam turbine principles been adapted long before this?" After many years of association with research and development, I have learned how difficult it can be to transfer technology into application; this remains one of the greatest challenges engineers face.

To understand the basis for space exploration, one must start with a recognition of rocket fundamentals. So much has been said about rockets and their use during the last two decades that it is tempting to skip over the subject with a comment like, "As everyone knows, rockets produce thrust by propelling hot gases out the rear of the vehicle." While this is true, how can anything so simple to say be so hard to do that it took centuries to apply? Perhaps a closer examination will show that implementing simple concepts is often a most sophisticated challenge.

For my birthday in 1953 my sister gave me a book by Arthur C. Clarke entitled *The Exploration of Space.* Clarke did an excellent job of explaining the mysteries of rockets to laypersons. At that time about 100 former German prisoners and a handful of American engineers were already beginning to get serious about developing rockets with capabilities beyond military applications, but my own fascination for space exploration was whetted by this book.

Several pages and sketches were devoted to the rocket principle. A man on a wheeled dolly with a stack of bricks was able to propel himself and the dolly by throwing bricks to the rear one at a time. Assuming the dolly to be rolling on a virtually frictionless surface, expelling the mass of a brick at a certain velocity imparted a reaction to the man, the dolly, and the remaining bricks that was admittedly less than but proportional to the velocity of the brick. From this analogy Clarke showed that it did not matter what happened to the bricks after they were thrown—their propulsion action occurred as they left the hand of the thrower. The fact that rocket propulsion is completely independent of any external medium, clearly the case for the thrown bricks, has always been one of the hardest things to understand. A second important point Clarke made was that as the pile of bricks on the cart became smaller and the vehicle thus became lighter, the velocity increment provided by each brick increased. This highlights the fact that in addition to the velocity increase, there is an increase in acceleration produced by a rocket as the propellant is expelled and the rocket weight decreases.

Clarke neatly and logically carried the analogy further until it is clear that the final speed is due to the cumulative effect as brick after brick is thrown. Each brick adds a small increment to the velocity that is dependent on the speed at which it is thrown; thus, the final velocity depends on this and on the quantity of bricks thrown out. By using numerical examples, Clarke developed the relationship between the mass and velocity of the bricks and the mass and velocity of the vehicle after all the bricks are thrown, taking into account the fact that the bricks on the dolly must be accelerated along with the man until they are all gone. From this we learn that for the final velocity of the vehicle to equal the thrown velocity of the bricks, the starting weight of the man and the dolly plus bricks must be 1.72 times the final weight after all the bricks are thrown.

But what if we want to go faster than the speed of each brick? Yes, there is an answer for that, too. By using the same relationship, Clarke calculated that we could achieve twice the speed of the bricks by making the load of bricks 6.4 times the final weight of the man plus the weight of the dolly, and 3 times the brick speed if the starting weight is 19 times the ending weight. This tremendous multiplication factor appears to place a sobering limit on the practical application of rocket technology. This in fact was the major deterrent encountered by early engineers; the gravity of Earth is such that no practical rocket could be conceived that would allow a single-stage rocket vehicle to escape from this field.

From the second law of physics expressed by Newton in the form, force = mass × acceleration, the equation for the final velocity that may be imparted by a rocket to a single-stage vehicle may be developed as a simple expression involving rocket exhaust velocity and the beginning and ending masses of the vehicle. Showing the expression in mathematical form helps to understand the key parameters and their simple relationship. For a rocket stage operating in an ideal environment, that is, having no restraining forces such as drag or gravity, the velocity it can achieve is a function of the exhaust velocity and the natural logarithm of its ratio of gross weight to empty weight is,

$$v = c \times \ln (W_{gross}/W_{empty})$$

where v is the rocket stage velocity, c is the exhaust velocity, ln () is the natural logarithm of the mass fraction term, and W is the weight of the stage either loaded with propellant (W_{gross}), or empty (W_{empty}).

The theoretical performance of rocket propellants can be defined more usefully in terms of the fuel specific impulse (I_{sp}), where I_{sp} = pounds of thrust per pound of fuel per second. Using this relationship and introducing the gravitational constant, g, for the pull of Earth's gravity produces an expression for what is termed "ideal velocity":

$$V_{ideal} = I_{sp} \times g \times \ln (W_{gross} / W_{empty})$$

This ideal velocity offers a simple way of comparing the potential of given rocket stages. It is primarily the propellant chemistry that determines the exhaust velocity or fuel specific impulse, although the efficiency of the rocket nozzle is also a factor. The V-2 and early liquid propellant rockets developed in the United States used ethyl alcohol and liquid oxygen as a propellant combination and produced I_{sp} values of about 240 seconds. Later, more energetic jet propulsion fuels were used instead of alcohol, and finally liquid hydrogen and liquid oxygen became standard, providing specific impulse values of about 450 seconds—almost twice those of the V-2.

The ratio of weights or the so-called "mass fraction" is dependent on structural efficiency and fuel-oxidizer densities. The matter of designing lightweight structures had long been a major challenge for aeronautical engineers; dealing with this issue for rockets required greater concern for materials able to withstand high temperatures, but was simply an extension of current thinking. Better cooling techniques, higher pressures, and better propellant pumps have improved the thrust of rockets such that their efficiencies, combined with the mass fractions available using existing materials, have greatly exceeded those of the early rockets.

While ideal velocity is useful for comparing the relative merits of rockets, the actual velocity achievable by a given stage has to account for three principal effects associated with the "real-world environment." These effects can be treated simply as subtractions from ideal velocity. The first is the effect of gravity; this is typified by the force during launch that is always pulling the rocket toward the center of the Earth; the second is the effect of drag as the vehicle passes through the atmosphere; and the third is due to atmospheric pressures acting on the nozzle to reduce rocket thrust. The last two are factors only during the boost phase in the atmosphere, but gravity effects are always present. In deep space, far from Earth, the Sun may produce the dominant "gravity" force, but such effects must always be reckoned with.

Expressing the burnout velocity for a single-stage rocket in simple terms:

$$V_{burnout} = V_{ideal} - dV_{gravity} - dV_{drag} - dV_{thrust}$$

where the dV terms represent incremental subtractions.

On a relative basis, the losses in velocity during a rocket launch to space caused by drag amount to only 5 percent or so of the required velocity. Thrust loss due to nozzle effects accounts for a similar percentage, depending somewhat on the optimization of nozzle design and staging altitude. The biggest losses are due to the pull of gravity; the effect of this reduction for a rocket rising vertically from the surface of Earth is 20 miles per hour for every second of climb—1200 miles per hour for every minute!

The prohibitive size of vehicles having a theoretical capability to escape from Earth led to the concept of staging. The idea was simply to stack two or more rockets so that the upper ones were treated as payload for the lower ones, with the advantage that the heavy structure of a lower stage could be discarded after fuel was expended and it had served its purpose. By starting over with a smaller rocket having an initial velocity equal to the final value for the previous stage, dead weight was carried no longer than necessary.

Reducing the weight of rocket vehicles offers such gains that many changes in the design of structures evolved from the baseline aircraft technologies. For example, pressurized stainless steel tanks with very thin walls were used on Atlas and other missiles. Like a balloon, these structures were stiff under pressure, but during their manufacture and handling they had to have hardback supports to maintain shape. The fact that empty weight is so important to rocket efficiency is still one of the principal reasons that rocket vehicles seem to operate on the ragged edge of failure. The luxury of large structural margins and redundant systems simply cannot be afforded if payload capability is maximized.

The notion of staging is taken for granted today; however, long after engineers began considering the matter of staging there was controversy about whether rocket staging would actually enable us to launch a meaningful payload into deep space. There are complications, of course; staging operations have often resulted in the frustrations of launch failures, but staging is now accepted as a way of life.

An associated problem for the early rocket missiles was that of warheads reentering the atmosphere at high speeds. The friction of the air at high speeds caused such severe heating that ordinary metal would simply burn in the atmosphere. The development of blunt entry shapes and ablative cooling techniques made this seemingly impossible requirement achievable.

Returning to the analogy of the ship and the rocket vehicle, it is appropriate to relate the two technologically as capable of carrying payloads for long distances. It is also appropriate to liken the compass, chronometer, and sextant that were required for successful voyages by ship to the guidance technologies required for predictable navigation in space. Attitude stabilization, always a requirement for celestial navigation, is a condition not achieved as readily in weightless space as on the seas. Not only was the attitude control of a spacecraft necessary for supporting celestial navigation, but in the same manner that the ship had to be pointed to take advantage of the wind and to make good the course desired, the spacecraft needed a stabilized platform to orient rockets for course corrections. Attitude stabilization was also needed for orientation of solar panels toward the Sun, for orientation of the high-gain antenna providing reception and transmission of low-power signals, and for pointing instrument sensors that were to serve as the eyes and ears of the spacecraft. An attitude control system, combined with the space equivalent of the chronometer and sextant, made it possible to determine the positions and trajectories of missiles. Doppler radar tracking systems became a better choice for tracking and guiding space missions, because transponders in the spacecraft were simpler and lighter than onboard position determination systems.

With continuous knowledge of position and some ability to control steering or midcourse rockets, the integration of the trajectory parameters could be achieved with the help of computers to keep track of position information. For early spacecraft it was better to have this integration of guidance information occur on the ground, because it was possible to use large, powerful computers that could not be carried into space to perform this function. Indeed, early tradeoff studies showed clearly that everything possible should be accomplished with equipment on the ground to save all the precious weight aboard the spacecraft for necessary components. Of course, this made the telecommunications link for commands critical; perhaps this departs from the analogy between the exploratory spacecraft and the early ocean explorers, who were completely out of touch with the world once they left the shore. Fortunately, telecommunications evolved along with rocket guidance and control technologies, such that radio transmissions over large distances were possible with low power.

In addition to these basic technologies essential to space exploration, the stimulus and motivation of man was required. In reviewing history, it seems that the technologies were often ready before this motivation occurred. Certainly in recent years this has been the case. While it is hard to say what actually started the space "snowball" rolling, the Russian plan to launch Sputnik into orbit clearly galvanized this country into action in the late 1950s. Not only did it force an appraisal of the state of technology, it also caused a coordinated look by American politicians, industrialists, and researchers at what the United States should do to achieve preeminence in space. Simply put, we entered a space race we perceived to be important based on the Russians' plans to launch a satellite into Earth orbit.

It did not take long to realize that we had the technologies in hand to begin such an effort. The books of Arthur Clarke and other science fiction writers started us thinking. Wernher von Braun wrote a series of articles for *The Saturday Evening Post* in which he described the various aspects of rocket propulsion and related technologies, and what could be done to put them together in a logical fashion for the exploration of space. His articles were based on sound engineering principles studied over the years, plus his strong belief that it was time to combine these technologies and to do some of the things that he saw were possible. His articles were timely and helped to convince a large segment of the population that such feats were not only possible, but that it was time to proceed.

Our defeat of the Germans and the spoils of war had left the United States with several partially completed V-2 vehicles and a large amount of information on the design and development of rockets. In 1948, I was working at North American Aviation in an engineering department concerned with trainers, fighters, and bombers when an opportunity arose for me to join a newly formed Aerophysics Department that had been assigned special studies of the V-2 technologies and their potential. Dale D. Myers, an aerodynamicist who had joined the aerophysics organization to head the new missile aerodynamics activities, offered me a chance to work in this new field. It was several months before I was able to complete an ongoing assignment and transfer, but early in 1949 I entered the fascinating world of missiles. My association with Dale was to continue on and off over many years, for he later became Apollo Program Manager for Rockwell International, and in 1970 we were reunited at NASA Headquarters when he became Associate Administrator for Manned Space Flight. Throughout my career his influence has been inspiring to me, for he always seemed to possess a rare combination of experience and insight needed to guide new technical efforts, and a gift for leading a team and getting the most from it.

Shortly after my transfer to the Aerophysics Department, work began on the design of a rocket-launched, winged cruise vehicle using a V-2 as the basis. Confiscated German rocket engines were available for tests, and facilities were built to improve them as preliminary design activities began. Needless to say, many "bootstrapping" studies were initiated, as we had to develop our own data base for dealing with supersonic flight, thermodynamic heating, and the new propulsion technologies.

A major program called Navaho evolved in 1951 from these early V-2 follow-on developments. It was defined as a rocket-launcher, ramjet cruise missile combination, to be ultimately capable of flying 5500 nautical miles. Simply stated, the Navaho program objective was:

> A ground-to-ground, guided missile, capable of carrying a heavy special warhead over a maximum range of 5500 nautical miles at supersonic speed (Mach No. 2.75 or higher), with a radius of error at the target of 1500 feet or less for 50 percent of the missiles launched.

At the time this challenging objective was formulated, there was less reason for optimism than when the similarly simple Apollo objective of sending men to the Moon was pronounced.

Navaho development was planned to be carried out in three phases. First, a prototype cruise missile powered by two large turbojets was to test the aerodynamics and flight operations. Following about a year later was to be a rocket booster/ramjet cruise missile combination capable of flights of about 2500 nautical miles to test the launch concept and the ramjet propulsion systems. The third phase was to be the operationally suitable weapon system with complete capability.

Requirements for this missile program included the development of the rockets, ramjets, structures, propellants, and tankage, as well as the high-technology guidance and control systems. Also included were the procedures for rocket launchings plus the development of launch facilities, telemetry, and tracking needed to accomplish the tests and operational checkout of all systems. During the next 6 years, a remarkable amount of progress was made toward these simply stated, but hard to achieve, objectives. Twenty-seven flights were made with the turbojet-powered vehicle designated the X-10, including supersonic flights to Mach 2.0. These did much to further the guided missile technologies for disciplines other than rocketry.

The rocket-launched, ramjet-powered Navahos were for many years the most impressive missiles to be seen at the Cape. Standing about 100 feet tall and weighing approximately 300 000 pounds, their 405 000-pound-thrust boosters were the most powerful in existence. Although they did not fare as well in flight as the horizontal takeoff, turbojet-powered X-10, many lessons were learned about the vagaries of rocket launches. When the base program and a flight extension finally concluded, nine rocket launches had been made, three of them followed by successful ramjet flight operations. The success ratio was not impressive, but after more experience with rocket launches, the record did not look as bad as it had seemed at first.

The now familiar concept of the launch complex with its distinctive gantry and blockhouse was not initially obvious for missile launches. Orion Reed was at the Cape from 1951 to the present, and as the base manager for North American during the Navaho flight test program he was involved in all the debates over how to provide for test operations, with due consideration for crew safety. He recalled that the one way to ensure safety from possible explosions was with separation distance; the price paid for this simplistic solution was long communications and data lines, plus inconvenience for access to the pad and for observing the equipment and operations. Television systems were not developed enough for widespread use in 1951, and it was essential to have direct viewing and ready access to the launch pad during the countdown.

The compromise struck for the Navaho launch site resulted in a small, hemispherical blockhouse built of sandbags and concrete that would house a few critical personnel during the final count and launch operation. Others were separated from the site with the long communications lines and compromised view of the launch. The small shelter was closer to the launch pad than the distance later chosen as a standard. The Air Force development of the pads, including 12 and 13, used for all the lunar and planetary launches on Atlas/Agena vehicles, were based on more detailed studies and were much more rugged. These blockhouses had roofs made of 20-foot-thick concrete, several feet of sand, with more concrete on top. They could house 50 to 100 persons and allowed indirect viewing of the launch site by periscope and closed-loop television.

Reed vividly remembered the blowup of an Atlas/Able vehicle during a static firing on September 25, 1959, shortly before the planned launch. The explosion and rain of debris was very close to the blockhouse, and the launch crew was grateful for the protection provided, since the pad was pretty much destroyed. Throughout his launch operations career Reed spent many days and nights in blockhouses (the 200th Atlas launched Ranger 6 in January 1964, and he was involved in most of the space missions launched by Atlases). He played a major role in helping to mature countdown and launch operations into a science.

The manner in which man and automated systems can work in partnership is illustrated by a solution to the problem of accurately guiding a Navaho test missile during approach and landing. In conjunction with an autopilot and inertial guidance system, a radar altimeter was used to flare the missile as it approached the runway on a predetermined glide slope. The accuracy of this automatic flare system was suitable for closed-loop operation, but the autopilot and digital navigation sensors available at the time were not capable of laterally aligning the missile flight path with the relatively narrow runway.

Orion Reed recalled that to achieve the necessary directional control for the touchdown and rollout phases, a simple optical tracking instrument was positioned at the far end of the runway with a means of generating an error signal that could be transferred by radio command to the missile autopilot. To operate the device, a man peering through the telescope kept crosshairs aligned on the nose of the missile, and the lateral deviation error signals were fed back to the missile autopilot for making the necessary heading corrections.

The system was used satisfactorily during flight tests. However, the optical device became affectionately known as a "hero" scope after a braking parachute failure allowed the missile to continue down the runway toward the hapless controller who was staring it in the eye as he guided it directly toward himself. A disaster was narrowly averted, but it became obvious that the person closing the lateral control loop was in jeopardy if landing overshoot occurred. For his benefit, a trench was dug near the instrument so that he could dive into it for protection if the need arose again.

At the time of the Navaho developments in the early 1950s, it did not appear that guidance and control systems using inertial platforms could perform on long-range missions without frequent updates. A system called the "stellar supervised platform" was developed to ensure that the drift of gyros was corrected by frequent star sighting inputs from the equivalent of a sextant used during ocean voyages. Concurrently, improvements were made in the performance of gyro systems, double integrating accelerometer systems, and other elements of basic attitude reference platforms.

As a result of this concentrated effort on guidance and control technologies, the capability needed for accurate intercontinental ballistic missile guidance systems became available. Ironically, the General Dynamics Corporation capitalized on these advances and began promoting the intercontinental ballistic missile (ICBM) concept called Atlas in basic competition with the Navaho. Thus the Atlas missile, a 1½-stage vehicle combining rocket engines, guidance and control concepts, pressurized stainless steel tanks, and other technologies developed at North American for the Navaho program, eventually became the promise that resulted in the demise of Navaho in 1957.

A host of developments that evolved during the Navaho program also provided a technology base and played major roles in space exploration. North American extended V-2 rocket engine technologies, and the Rocketdyne Division later produced rockets for the Thor, Jupiter, Atlas, and Saturn vehicles, for the Apollo spacecraft, and for the Space Shuttle. An Autonetics Division provided guidance and control capabilities, becoming heavily involved in upgrading the guidance and navigation technologies for ICBMs and space vehicles. The Missile Development Division developed aerothermodynamics concepts, structure capabilities using high-temperature materials such as stainless steel and titanium, design and manufacturing technologies such as diffusion bonding and chem-milling, and a well-trained cadre of engineering and management talent that designed and produced the Saturn launch vehicles, Apollo spacecraft, and Space Shuttle vehicles.

The cancellation of the Navaho program and the successful orbiting of Sputnik 1 were, to my career, closely coupled shocks, like the double impact of a sonic boom. For the engineers who were laid off because of the Navaho cancellation, and for those of us with supervisory responsibilities who had to decide which friends and associates would stay or go, it was a dismal period. I was very glad when the chore was over and those of us remaining were assigned to study the use of Navaho technologies for space missions.

Although many were applicable, it was the rocket boosters that gave us the immediate capability to prepare for traveling into space.

Even with the significant developments spawned by Navaho, there was one major shortfall in our ability to design space missions. Our missiles had all been governed by flight in the atmosphere, where aerodynamics was the dominant discipline. One might think that going into airless space, where the principal forces are caused by the gravitational attraction of bodies should have been simpler, but we did not have in hand the parameters defining gravitational forces needed to determine space trajectories. We also lacked the basic equations and the programming to integrate trajectories; besides that, our computer was very large in size and very small in capability.

Encouraged by my superiors to find help, I visited Professor Seth Nicholson of CalTech and also discussed our needs with astronomers at the Mount Wilson and Palomar Observatories. While doing this, I met a retired astronomer, G. M. Bower, who had calculated the mass and orbit of Pluto. Although he had done most of his integrations using mechanical calculators, Bower had some recent experience adapting the equations to computer language. He joined our small group and helped generate the so-called "n-body" equations used in computer computations of space trajectories.

Further evidence of our lack of capability to compute trajectories became painfully obvious when I contacted G. M. Clemmence, then head of the Naval Observatory, to obtain ephemerides for the Moon, Mars, and Venus. Values of the orbital parameters for these bodies were essential to navigation, and the Naval Observatory was the central repository for such information. Clemmence surprised me when he stated that he could provide computer input data suitable for computing trajectories to Mars and Venus, but that data were not available for developing trajectories to the Moon. The reason given was that many variables, such as Earth's tides, affected the Moon's orbital path, so that it was not easy to exactly predict long-term values for the Moon's whereabouts. This revelation begged the obvious question: if we don't know where the Moon is going to be when we launch, how can we determine in advance how to get there?

This kind of activity was not unique to North American. All over the country aero industry teams and research groups were doing the same things, with perhaps one of the most notable efforts led by C.R. "Johnny" Gates at JPL. He and a small group in the Systems Division developed integration methods, adapted them to computer operations, and soon began to put their knowledge to work on real projects. Within a brief period, nearly everyone learned how to compute space trajectories. A new discipline, called astrodynamics, sprang out of the combined aerodynamics and celestial mechanics backgrounds of aeronautical engineers and astronomers. Computer technologies were also driven hard by the obvious need for larger matrices and faster operations.

Once staging concepts began to be generally better understood, the multiplication factors for upper stages were simply reduced to engineering terms. Initially, staging had been thought of as launching one rocket with one or two others on top, to be ignited sequentially as soon as the prior stage had burned out. Now coast periods between firings were being used to provide more control over trajectories. Coast periods between stages could be accomplished with no thrust at high altitudes; the fact that the stages remained connected had no significant impact on performance when the vehicle was not in the atmosphere and being affected by drag.

For example, in the case of synchronous satellites, which had to orbit at 23 000 miles above Earth, coasting up to synchronous altitude and then firing the last stage to produce the right velocity for staying at that altitude allowed the satellite to be put into a precise circular orbit. Coasting trajectories became known as "parking" orbits and were frequently used to launch vehicles into deep space from a position other than the launch site. Such staging considerations offered many possibilities for tradeoffs; the precision timing required for leaving the launch pad was reduced, because it was possible for the vehicle to coast part way around Earth—or even to make an orbit or more—before the next stage was fired.

The payloads for early missiles were warheads, which usually were inert until carried to the target site. Thus, from the standpoint of integration with the vehicle, they merely weighed so much, were so big, and otherwise had a modest interaction with the design of the vehicle. As the aerodynamic shape of the warheads was especially critical to the reentry thermodynamics of missiles, much work was done in the development of ICBMs to solve the thermal and aerodynamic problems of reentry. In addition, knowledge of high-temperature phenomena was needed, and materials able to withstand high temperatures had to be developed. These were especially important for missions that required reentry to Earth's surface or entry into the atmosphere of Mars or other planets. Of course, we were not concerned about entry aerodynamics for the first planetary exploring machines because they were merely one-way vehicles that flew by or in the vicinity of planets. Nor did we need such technology for spacecraft designed for landing on the Moon, where there is no atmosphere.

One of the dramatic changes during the early years of transition from missile launches to space launches was caused by the change from passive warheads or payloads having very few active elements to what became known as spacecraft, which were vehicles in their own right. The effects of this transition became painfully evident during the early space launches at Cape Canaveral: those with experience launching missiles tended to think of the principal process as readying the rocket vehicles, launching them, and tracking them into space into prescribed orbits. After a few missions in which the launch was over within minutes and spacecraft operations became long-term tasks, the realization dawned that what had once been prime aspects of missilery were now relegated to support roles. Certainly, launch vehicles and launch operations were no less important; but now, launching the spacecraft at the proper time into the proper orbit merely allowed the spacecraft to get on with the real job of exploring. There is no obvious analogy to this with the early days of ocean exploration, since the beloved ships were "single-stage vehicles" that not only carried the explorers from shore to shore, but also brought them home. Rocket launches, even for boosters employing three stages, are over quickly relative to the long journeys of spacecraft; after launch they simply become "spent vehicles" that serve no further purpose.

As spacecraft became more than inert payloads, further evolution of rocket vehicle technology was required. Propulsion systems now had to be stored in space for long periods of time and operated remotely after exposure to the vacuum and thermal radiation of the space environment. Attitude control systems for launch vehicles had to work for only a few minutes; thus the drift rates and wear problems associated with short-lived missiles were completely different from those expected of spacecraft, which had to spend months in orbit. The guidance and control systems necessary for accurate midcourse corrections, terminal maneuvers, and other functions required precision and updating of position so that after months in orbit or interplanetary space, exact pointing of the rocket motor or aiming of the instruments would be possible. While the basic technologies were similar to those required for launch vehicles, the demands for precision, for miniaturization, and for long-life operation in a somewhat hostile environment were greater.

Vehicles designed to land on the Moon or the planets required retro rockets to decelerate the spacecraft for landing. In principle, retro rockets changed the spacecraft velocity in the same way as booster rockets launched from Earth, except that the velocity increments they provided were used to reduce the velocity from high beginning values to zero at the point of touchdown. This led to the use of an analogy as a basis for determining the probability of mission success in planning considerations for lunar landing spacecraft such as Surveyor. A study I made in 1962 devoted considerable thought to this matter and resulted in a paper, replete with statistical probability curves, entitled "Probable Returns of Present Lunar Programs." Because this analysis offered significant possibilities for misuse by critics, it was stamped *For Office Use Only* and had a limited distribution. While soundly based on existing launch vehicle statistics, the probabilities of success using statistical data available from launch vehicle experience showed that less than one out of three flights aimed toward landing on the Moon could be expected to be successful.

The lunar landing/launch vehicle analogy became useful for illustrating the combination of technologies involved and the engineering challenges that had to be addressed for such missions. Actually the landing is the reverse of the launch in sequence, but a surprising number of the steps are analogous. The simple diagram from the 1962 study is reproduced here along with an explanation of its meaning.

A launch operation starts with a zero velocity as the vehicle is sitting on the pad. At the end of the launch, injection into orbit allows for some variation in the firing accuracy from the early stages; adjustments by a vernier engine make up for any deficits or excesses in velocity, orientation, or position in space. In contrast, the lunar landing vehicle begins its terminal maneuver with some finite but uncertain velocity and must arrive at the surface with zero velocity after a 240 000-mile trip taking some 90 hours. The landing would not be successful with any sizable horizontal or vertical velocity components at the point of touchdown, for the spacecraft would either tip over and be useless or be destroyed by the crash.

Other similarities and differences are highlighted by comparisons in the simple diagram. There are several more steps involved in a spacecraft landing mission, not to mention the fact that a landing attempt is not even a possibility until after a successful launch has been achieved.

Following the separation of the spacecraft from the upper stage of the launch vehicle, attitude orientation is needed to point the solar panels toward the Sun and the antenna toward Earth. This is the cruise mode, which continues until the time of midcourse correction. The velocity change for the midcourse correction is determined using computers on the ground, and commands are loaded into the spacecraft to properly orient the thrust axis and to fire the motor for the time needed to make the desired trajectory correction. This series of maneuvers in the midcourse involves some risk, because it is necessary to turn the craft away from the Sun-Earth orientation to a given inertial attitude, to fire the vernier rockets for a fixed amount of time, and then to shut off the rockets and return the craft to the cruise attitude, again acquiring the Sun and Earth.

As the spacecraft approaches the Moon at a given height above the surface (this must be fairly accurately determined by a triggering radar), the large retro motor must be ignited. For Surveyor, a highly refined solid rocket motor of spherical shape was used. When it was built, it had the highest performance in terms of mass ratio and specific impulse of any solid rocket in existence—Surveyor was its first space application and true test. After the burnout of this motor, it was essential that the spent rocket be separated and that staging occur in a manner that did not tip the spacecraft or cause it to lose attitude control. The retro motor then fell to the Moon, ahead of the spacecraft, while the vernier engines on the spacecraft slowed it to further reduce the velocity of approach.

A closed-loop radar system was used to guide the spacecraft down to the surface. Engineering for this system presented challenging difficulties, partly because we lacked detailed information about the surface of the Moon; thus, its radar-reflective properties were only speculated on the basis of engineering models. Another unknown at the time was the interaction of the radar system and the tenuous atmosphere created by rocket exhaust, possibly causing undesirable radar dynamics. There simply was no good way of testing these environmental combinations prior to the first Surveyor mission.

The vernier rockets used to reduce the remaining velocity of the spacecraft to near zero at touchdown were throttleable liquid propellant engines. Throttling was not, at the time of Surveyor, a common practice on liquid rockets; this vernier system was specially developed. Determining the vertical approach velocity with radar seemed relatively straightforward; however, determining the horizontal velocity component, which was just as important, was not so easy. With the small radar baseline on the spacecraft it was not possible to track the horizontal velocity until the craft was very close to the Moon. This meant that the buildup of horizontal velocity during the

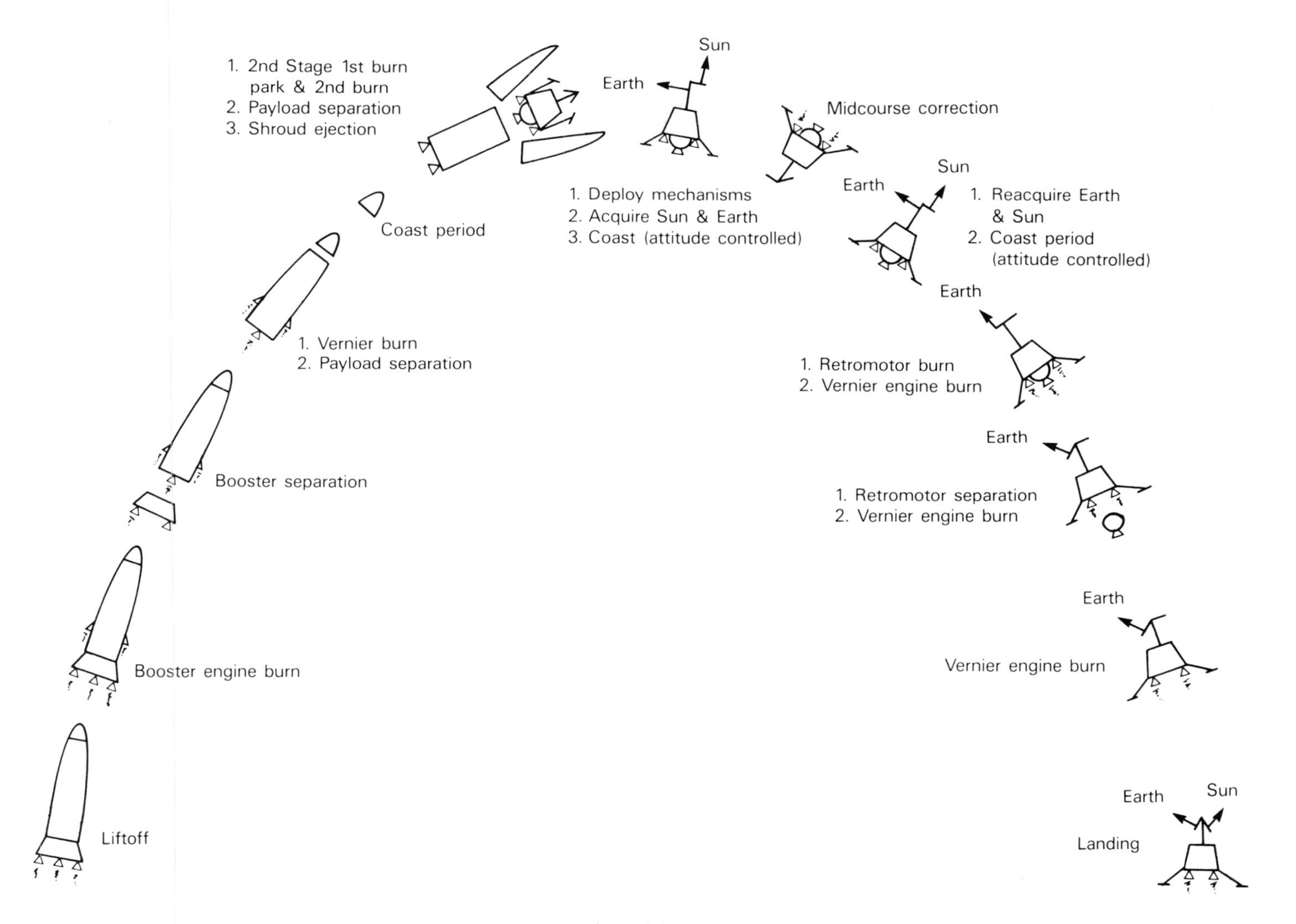

Launch and landing

retro motor burn, separation, and approach had to remain within finite limits in order for the lateral correction capability of the swiveling vernier rockets to suffice. Taking into account all these functions required of a lunar landing system, the probabilities of success were calculated based on existing launch vehicle statistics. These indicated that the landing itself was a fairly risky proposition, with a probability of success far less than 50 percent.

As mentioned earlier, a concept accepted at the time was that launch vehicles undergoing development should experience 10 development flights before being used to carry out operational missions. Had we applied this concept to Surveyor landing systems and taken the 40- to 60-percent reliability of existing launch vehicles into account, it would obviously have taken many launches to enable us to develop and check out a suitable Surveyor lander. These discouraging statistics increased our concern for the thoroughness required during Surveyor research and development activities, but they were helpful to the planning process within the program office and resulted in the adoption of the multiple-spacecraft block concept, calling for a minimum of three-of-a-kind for most configurations. The idea worked, except for the second block of Rangers, which failed to achieve a single success.

Perhaps the greatest driving force in the evolution of spacecraft was the challenge of providing "long-life" capabilities for systems that had to operate in a hostile environment with minimum human interaction, and that required very modest amounts of weight and power. After seeing a number of different spacecraft concepts developed, I have come to the conclusion that having to design and build to severe constraints actually improves the evolution process. When engineers have ample amounts of weight, power, and other resources to start with, they almost immediately expand their desires to exceed those capabilities and develop self-imposed problems from trying to juggle all the "what ifs" and "druthers" into something real. They usually make much more work for themselves, and in many cases design less suitable systems as a result. On the other hand, I have seen designs evolve when severe constraints were imposed, requiring single-purpose objectives and simple, direct applications of basic physics, that resulted in the most clever advances in technology to perform the necessary functions.

As the technologies employed in launch vehicles and spacecraft have become more complex, the number of engineering man-hours involved in design and development have increased. I asked Dale Myers to discuss the changes he had seen in the process and the reasons for them. He did not give a pat answer, but offered observations from his own career to support the changes in engineering effort versus production rates. When he started at North American in 1943, the factory was producing 25 Mustang fighter airplanes a day. In the early 1960s he was in charge of the Hound Dog missile program, which produced 20 missiles a month. During the final period of the production of Apollo Command and Service Modules, the production rate was 6 per year, and when he was responsible for the B-1 bomber effort in 1974, the bombers were being produced at a rate of 1 every 2 years. The escalation of effort has also severely increased the cost per pound of hardware, making the problems of estimating program costs much harder for space vehicle planners. However, that is another story.

No matter whether you are a launch vehicle proponent or a spacecraft engineer, it is obvious that rockets have provided the key to the exploration and exploitation of space. At times it appears that we have lost sight of their importance in the scheme of things; we seem to be complacent about the potential gains that might accrue from continued emphasis on their improvement. Far-out concepts for doubling or tripling their efficiencies are barely being researched, if considered at all. Are we once again experiencing that lag in the engineering advances of existing technologies until necessity, not opportunity, becomes the "mother of invention"? Time will tell.

A Million Things To Do

Planetary exploration is an ancient and sanctified pursuit, underway, if you count exploration of Earth, for at least a million years. In spite of all the years, we have hardly begun the task of exploring Earth. There are great areas of land that have only been sampled, and we have only the skimpiest knowlege of the 69 percent of the surface that is under water. It was only after we left Earth for orbit that we began to understand our planet better, recognizing it as one of the most varied and fortunately endowed objects in the solar system. During the last two decades we have learned much by traveling around it, walking around the home block for the first time, so to speak. Views from the vantage of near-Earth space have provided exciting new perspectives of the planet, clearly revealing the land masses and their encircling oceans, the continent-sized storm formations, the restless clouds, and the dynamic environment in which our lives are spent. Being alive, we care about life, and for this our magical and unique atmosphere is the key. Using sounding rockets and satellites we have plumbed and defined this thin planetary coating that shields us from high-speed particles, filters out lethal radiation, ameliorates insupportable temperature variations, and carries to us the water without which life would soon vanish.

In these same decades we have also gained insight into our planetary neighbors. Using increasing sophistication in our automated spacecraft, we have visited the Moon, Venus, Mercury, Mars, and those strange gas giants, Jupiter and Saturn. Close-up photography has made details of the Moon's face familiar, its scarred surface recording both great eruptions from within and violent bombardment from space. Cataclysmic history as preserved in scrambled detail on that pockmarked surface is being used by scientists to fill gaps in Earth's remote past, for similar scars on our own planet have been all but obliterated. Planetologists see the Moon as a Rosetta stone for our solar system, helping to unlock the cyphers of an enigmatic past.

Men first began to explore Earth by looking at their immediate surroundings, gradually widening their travels to encompass larger areas. With our new exploring tools, we have a better way. Obviously, an unknown planet should not first be examined locally, with a restricted view, but should be seen in its entirety, from flyby and orbit. We should begin with reconnaissance of gross features: clouds, continents, polar caps, mountains. We should examine the atmosphere, if any, as knowledge of the atmosphere not only offers many answers about the nature of the planet and its history, but also prepares for the engineering of successful landings.

The next phase should involve landings at preselected sites, chosen as the likeliest to provide a maximum of information. These are often at the boundaries between different types of terrain: at the edges of polar caps, the bottoms of large hills or valleys, and the rims of high plateaus affording distant observations. Finally—before the risky and costly landing of men—we might use automated roving or flying vehicles, remotely guided from Earth, affording the intelligence-gathering skills of our best substitutes for human eyes and human senses without risk to life. Then, if conditions warrant, we should send a combination of machines and the best multisensory decision-making resources we have—human beings—when the extra costs of life support and confident retrievability can be justified.

After a few missions are flown, some critical components are more or less taken for granted and put from our minds because they can be counted on. Persons coming into established projects or those observing projects underway may never realize how everything came together the first time or have a good perspective for judging the totality of the logic and the processes involved. No major technical undertaking is ever done from scratch—we have mentioned the tremendous contribution to space missions of missile programs—but developing appropriate plans and putting the necessary systems in place to support the first Mariner flights took a great deal of ingenuity and effort.

As mentioned repeatedly, the biggest initial hurdle to exploring other planets was our marginal ability to escape from Earth. The launch vehicle has always been a limiting factor, restricting the mass of spacecraft to a degree that has challenged designers to provide useful payloads. This constraint figured in making the Moon our first target; however, it would have been foolish to bypass the Moon, the nearest neighbor of Earth, to go first to far planets. Yet there were compelling scientific reasons to obtain information about several planets as quickly as possible; by studying them we obtain perspective about their similarities and dissimilarities to Earth, providing advances in knowledge of the solar system through the use of broadly based comparisons.

Overall plans for programs were developed in the Lunar and Planetary Programs Office and reviewed in a number of forums. Reviews were provided by the scientific community, often through the Space Science Board, set up for that purpose by the National Academy of Sciences. During the budgetary approval process, mission proposals passed through the NASA administration and then to the committees and bodies of Congress, providing repeated opportunities for us to examine and defend our rationale for exploring the Moon and the planets.

A key step was the selection of payloads, requiring not only scientific review of the entire mission, but also the allocation of priorities among individual experiments. I believe that the process NASA developed for payload selection has withstood the test of time. It required a subtle yet complex interaction among people, machines, budgets, and politics (after all, these were publicly funded programs). Also intrinsic was a tolerance for considering everything of relevance, from the engineering of a tiny subsystem to testing a theory about cosmic origin.

From a scientific viewpoint, the most difficult part of planning missions was choosing experiments that addressed the most fundamentally important questions. After lengthy discussions, order finally was provided for the process through the coalescing of views on major classes of questions for the Moon and the planets. Once these had been defined and accepted, it was possible to develop balanced experiment packages and to consider individual experiments in proper relationship to others. The simplifying approach also helped address experiment sequencing or priority questions; in some cases time-critical interactions with other experiments affected priorities for instrument selection.

Four classes of scientific experiments were initially defined to address major planetary questions; these are now logically explained by thinking about examining a planet "from the outside in."

The first class of experiments addresses a planet's environment as determined by external influences such as the Sun, especially radiation, particle fluxes, or varying energy fields and their effects. Planets in our solar system exist in an environment largely Sun dominated, although the environs of a planet can be modified by the presence of its own unique magnetic field.

Field strength and orientation also provide insight into many characteristics related to the origin and nature of the body. The name given the first class of experiments was simple and descriptive: particles and fields.

The second class of experiments deals with planetary atmospheres: their compositions, densities, pressures, temperatures, clouds, or other peculiar features. Instruments for this class of experiments come in a variety of forms, ranging from direct sampling to remote sensing.

A third class of experiments known as body properties is broadly directed at the celestial body itself: external shape, mass, density, precise orbit, and rotation rate and direction. There is interest, too, in mass distribution and tectonic condition (whether the planet is volcanically active or quiescent), because these conditions can be related to planetary history and help in hypothesizing the conditions of evolution. Experiments in the third and fourth classes generally were labeled "planetology" experiments, a term coined to convey their relationship to the field of geology, but having a broader connotation.

The fourth class of experiments deals with the planetary surface: its texture, features, and composition. This includes topography—mountains, hills, valleys, craters, and other forms of nonuniform disturbance—and chemical composition and materials properties. It is of vital interest to determine whether the materials composing the planet are minerals like those on Earth or unique constituents. Physical measurements of surfaces—hard or soft, sandy or dusty, lava-like or deposited in other ways—are needed, as are measurements of other properties such as conductivity, temperature range, and magnetic susceptibility. Some of these characteristics call for in situ analysis, and a few require sample return to laboratories on Earth. Initially this class covered the broad questions related to the search for life; it was not until biology experiments like those carried years later by Viking that a special biology class was added to the four basic classes.

When all proposals for experiments were fitted into this framework, choosing balanced instrument payloads became easier. Looking back with the assurance provided by time and experience, I wonder why the simple process of defining major classes of experiments was so momentous; it does not seem very profound today. Perhaps it is because at the time no one had the interdisciplinary knowledge necessary to define the broad options of first-time missions. I remember listening to the individual scientists lobbying for their particular experiments, fearing that we might mistakenly succumb to "squeaky wheel" pressures and overlook a prime question that no one had fully considered. Once the four classes had been tested and accepted by most of the experiment proposers as encompassing, however, our confidence grew, and the concept of using them to develop balanced payloads was not challenged.

While working in industry on the design, development, and production of flight hardware, I had become accustomed to a "projectized" organization with strong, almost military, hierarchical leadership. I naturally supposed that it would be necessary to organize the total NASA effort, including the scientific component, along those lines if we were to successfully conduct projects with such demanding hardware requirements, engineering interactions, and strict deadlines. To my dismay, wiser heads than mine decided that the scientists who participated in space missions should be allowed to remain in their laboratories and classrooms, retaining as much freedom and independence as possible. I had grave doubts that we could make successful teams for the difficult missions unless the generally undisciplined scientists could be brought together with the other project members under rigid controls. This view was proven wrong, for the system worked and schedules were met most of the time. Now I realize that if my management concept had been forced on academic investigators, it would have severely limited the long-term dividends of space science, because it would have compromised the "fresh-eyes" benefit of their participation.

The NASA Space Sciences Steering Committee was a particularly important element in the total process. This committee was developed by Homer E. Newell, who had been a successful administrator of scientific activities with the Naval Research Laboratory before joining NASA. The committee as prescribed embodied an exceptionally balanced blend of managerial, engineering, and scientific viewpoints. It was chaired by Newell, who, as Associate Administrator, had the highest line responsibility within NASA for space science and applications programs. His alternate was the Chief Scientist, who, by organizational assignment, was necessarily concerned with the relationship between NASA and the scientific community. Other members of the steering committee were space science program office directors and their deputies—responsible for physics and astronomy programs, lunar and planetary programs, and bioscience programs. This body of eight had review responsibility for all scientific payload recommendations, with Newell as final selecting authority.

Within the steering committee a system of subcommittees was established, oriented along scientific discipline lines and chaired by NASA personnel, with members chosen for their specific scientific expertise. In the early 1960s seven of these subcommittees were appointed by Newell, covering the areas of particles and fields, solar physics, planetology, planetary atmospheres, biosciences, physics, and astronomy. Though the pyramid of committee and subcommittees may appear rather complicated and unwieldy, the various bodies interacted well and functioned smoothly, thanks primarily to the talented and dedicated professionals who served on them.

Whenever a major mission or series of missions was being planned, an announcement of flight opportunities (called an AFO) was made to the scientific community. This announcement outlined the nature of the mission, the types of experiments of general interest, and gave, to the extent possible, broad guidelines for proposed experiments. Proposals for specific experiments came from all quarters and were categorized and submitted to a subcommittee for review . Sometimes as many as 60 or 70 experiments were proposed for a mission that could accept only 5 or 6. Subcommittee responsibilities involved review to determine the scientific importance of each experiment, an assessment of its readiness to be integrated into the spacecraft, and an assessment of the competence of the investigative team. After subcommittee consideration, proposed experiments were placed into one of four categories and presented to the steering committee:

Category 1 consisted of high-priority scientific experiments that appeared ready to fly.

Category 2 consisted of high-priority scientific experiments that did not seem well matched to the mission or that might depend on technical developments not yet in hand.

Category 3 included promising experiments that might, perhaps with concerted effort, be prepared in time for the mission.

Category 4 comprised unsuitable proposals that, for either scientific or technical reasons, were deemed not appropriate for the mission.

This sorting process afforded a thorough review, yet left final selection to the management team responsible for making the mission as worthwhile as possible. It afforded ample opportunity for inputs from all sources and, in general, withstood the fairness test quite well. There were a few cases in which our selection process was criticized, but it was broadly accepted by the scientific and engineering communities as a reasonable approach.

Although a full-fledged member of the Space Sciences Steering Committee, I was in the minority as an engineer, along with Jesse Mitchell, Director of Physics and Astronomy. It was always a serious matter to select a specific set of instruments on a scientific basis and to determine how the meager amounts of weight, power, and physical space could best be allotted. This called for close collaboration between scientists and engineers in a manner that tried to account for all the variables. Though outnumbered six to two by scientists, I always felt that final decisions for payload selection gave ample consideration to the engineering judgments Jesse and I frequently brought into the overall process.

One thing that did puzzle me during the early payload discussions with scientists was an initial reluctance to fly camera equipment. Somehow pictures were not thought of as scientific, not as informative, for example, as a telemetered record of a varying voltage. A camera was not regarded as a scientific instrument at first, and to go after "picture postcards" was seen by some as an unscientific stunt.

This was a curious bias. Some of it may have been due to the fact that many of those hardworking scientists who were prepared for space science experiments had been accustomed to using only numerical data; thus, the fact that images represented a means of packaging data was not immediately obvious. In addition, some resistance might have been rooted in a wariness toward NASA and the motives of its administrators. Since NASA was a government agency dependent on popular and legislative support, some scientists may have suspected that we wanted photographs primarily as Barnumesque publicity attractions.

Charles P. Sonett, my deputy at the time, recalls one space science subcommittee meeting in 1962 involving more than a score of scientists gathered to consider the instruments to be flown on a future lunar mission, perhaps Surveyor. Most of the group, which included Nobel Prize winners, voted against flying TV cameras. Cameras had been supported only by Sonett (he was chairman of the meeting), and by Gerard Kuiper, a celebrated astronomer concerned with obtaining detailed images similar to those he was accustomed to seeing with telescopes. Sonett recalls that he reported this strongly biased view against flying cameras to me and I replied, "Fine, so let's fly them." I don't remember our dialogue in detail, but we did fly cameras on Rangers (and Surveyors and Lunar Orbiters as well), with broad agreement among scientists later that the cameras did much to enrich those spectacular missions.

In time the anti-image prejudice dimmed. It did not disappear overnight, but the startling effect of Mariner 4's shadowy images of craters on Mars and the torrent of images returned by the Surveyors did much to quiet the skeptics. I knew the issue was passé during an early Surveyor mission when I came upon Gene Shoemaker, a geologist of great foresight and conviction, surrounded by inquisitive colleagues from other scientific disciplines in a back room at JPL. They were prying from him interpretations of a newly acquired sea of prints, almost like zealots seeking meaning from a disciple reading the scriptures.

Nature imposed her own unappealable constraints on lunar and planetary programs through the geometry of the orbits of the Moon and the planets and their relationships to Earth and the Sun. Not the least annoying of the variables Mother Nature controlled was the weather, often a troubling factor at the time of launch and once even upon arrival at Mars. Mariner 9 had to orbit around Mars for quite a while waiting for a planetwide dust storm to subside. Weather here on Earth was more often a problem; for example, dense rainfall at Earth stations sometimes impaired the quality of returned data.

This snarl of planning variables was particularly challenging for Mars and Venus, because usable launch opportunities occurred only once about every 2 years. The fixed launch period scheduling problem caused the greatest consternation to scientists preparing their experiments. The entire schedule—planning, budgeting, development, and testing—had to be worked out backwards and events had to be time-phased so that they were completed when the launch period arrived. Launch opportunities typically lasted only about 1 month; when two launches were planned, they had to be made in rapid succession, often no more than 3 to 4 weeks apart. This in turn necessitated either dual launch pads or an extremely rapid turnaround and closely integrated use of launch facilities.

For NASA Headquarters managers a recurrent headache was the need to synchronize mission planning with congressional budget cycles: a desirable condition that seemed rarely to happen. Neither the actions of Congress nor the movements of the planets could be made to accommodate the other, and it was our task to do all the adapting that was needed.

As a practical matter, it was not reasonable to expect most scientific mission objectives to be accomplished with a single flight. Because of the unreliability of launch vehicles and the unrevealed problems of new spacecraft, it was difficult to know how many missions of a like kind should be planned, since there was no way of knowing which would succeed and which might fail. Scientists risked severe frustration by trying to perform experiments on flights for which the launching rockets and spacecraft were themselves largely experimental. Unfortunately, multipurpose flights, for example, developmental and mission-oriented flights, were a way of life. Even with considerable flight experience, the reliability of large multistage vehicles was rarely greater than 70 percent and often far less.

Many debates were devoted to possible ways to plan programs that might achieve flight objectives. Some thought it desirable to plan a series of test flights solely to develop the vehicle and the spacecraft, and then add the scientific payload. Others argued that even test flights should reckon with the possibility of total success. Another often-asked question was, should a series of missions have identical payloads so that those on failed missions could be duplicated as soon as possible, or should they be varied? The choices could work hardship to the edge of cruelty on individual scientists, for those whose experiments were not chosen to fly might have to wait for several launches, perhaps 4 to 6 years, before their instruments could be incorporated. In some cases experiments that could have been significant never got a chance because of vehicle or spacecraft failures. These were worst cases, though. Looking back, I am amazed that final results seem logically sequenced, as if we had had better knowledge of what to expect than we actually did.

If all went well and the spacecraft arrived at its target planet, the hour of the scientist was at hand. Data would come to Pasadena for sorting and for preliminary calibration and computing. The data alone were not enough; one also needed to know the spacecraft location and attitude and the times at which the data were acquired—all essential to the scientists but not directly within their control. It was then incumbent on scientific investigators to study, analyze, and interpret their results. Typically, the results would be published in the leading journals of the disciplines concerned; frequently NASA would prepare a special publication describing the mission and its results; sometimes a symposium would be held in which individual scientists would compare data and defend their interpretations. After the long, tense years that had gone into planning and executing a mission, I was always delighted to observe the cooperative and coordinated way in which highly individualistic scientists contributed to the common store of human knowledge.

It wasn't all roses, of course. The amount of time and work, the competitive environment, and the chanciness of investing peak career years in an unpredictable venture meant that some were inevitably disappointed. A few complained that the work of preparing proposals and then instruments and the difficulties in meeting strict schedules and complicated integration procedures were so overwhelming that their time was better devoted to laboratory work or other activities over which they had more control. And so it may have been—for them. But for those who braved the difficulties, waited for the opportunities, and sweated out risks of nonselection and mission problems to ultimately derive important new knowledge, it was a thoroughly worthwhile endeavor. Many scientists who worked hard on these investigations, and some who burned themselves out trying, feel that they received the greatest rewards of a professional lifetime for their efforts. Many, and perhaps most, felt that it was worth whatever it cost.

Quite apart from the planning and scientific processes of payload selection and development, there were many engineering and support functions to plan and prepare. Even with our missile background, there were few system test facilities that could simulate the space environment. Some vibration equipment useful for simulating aspects of the launch environment existed, but it was meager and limited in capability. Considering the high cost of a mission and the infrequency of planetary launch opportunities, it was important to check everything scrupulously while the spacecraft was still on the ground. In the moment of truth at liftoff, everyone concerned with a mission was prey to "for-want-of-a-nail" anxiety.

In the beginning, facilities were perilously jury-rigged or patched up. In 1960 spacecraft were assembled and tested at JPL in a building left over from previous missile work for the Army. Also used was a small building next door housing a makeshift shake table and small vacuum chamber. Within 2 years contracts were let and construction was begun on better facilities, including a realistic thermal vacuum testing simulator, but the new gear was not ready for use until five Rangers and two Mariners had been launched. Doubling up in the use of these limited facilities and extrapolating conditions well beyond known capabilities was a way of life until better facilities were completed.

Scheduling the use of available missile launch pads and blockhouses was continuously bothersome. Usually long lead times were required for preparation at the site, and uncertainties regularly arose about delivery and checkout of essential equipment. On top of these problems, uncertain weather conditions and unforeseen difficulties were likely to arise during intervening launches. As a practical matter, long-range scheduling always had to be iterated in the shorter term because of difficulties and program delays. Following an exact timetable was rare, and it was no less worrisome when things seemed to be going well.

Factors of a different sort that had a gross effect on planning were the availability of manpower and the status of funding. Manpower and dollars were not only essential in and of themselves, but the rate at which they could be applied influenced the development schedule and the scope of effort. Juggling these factors was a major management challenge; to decide the best ways to spend either resource involved tradeoffs that ultimately affected mission results.

In the early days of 1960–1962, when Ranger and Mariner were under development, many engineering practices were outlined that later became almost standard. For example, three nonflying Rangers were built to validate the final model: a spacecraft mockup, a thermal control model, and a proof test model. The first was used to confirm mechanical aspects of the spacecraft, including fits and clearances, cabling harnesses, and layout of equipment. (Even gifted designers and engineers can benefit from insurance against momentary spells of inexplicable oversight.) The thermal control model was suspended in a small, early-model vacuum chamber and subjected to the vacuum and simulated solar heating it would encounter in space. The uncertainties were large in those days, and for Mariner 2, engineers underestimated the heat encountered on the path to Venus. This resulted in the spacecraft running a high fever at encounter and dying of it a few days later. The proof test model, known as the PTM, was as similar to the flight spacecraft as possible. It was subjected to vibration and other tests somewhat above the actual predicted levels in order to provide suitable margins against the unforeseen.

The process of building additional vehicles for test purposes evolved through many variations as time went on. The practice had the additional advantage of providing a spare spacecraft in case of trouble with the prime article, and it gave us a duplicate to study on Earth if telemetry reported puzzling misbehavior millions of miles away.

Considerable debate was devoted to questions of the best testing doctrine. One view was that testing should not be performed on hardware actually to be used in flight because the stresses of testing might wear things out and would obviously affect equipment life. Another viewpoint was that flight hardware should be designed with sufficient margins to withstand both the rigors of testing and the mission itself, for this would eliminate weak links that could cause early failure.

An issue that to this day remains a matter of choice was highlighted by strong opposing positions taken by JPL and Goddard engineers when they were competing for planetary projects. The crux of the debate centered around the fact that spacecraft were going to operate "hands off" in space. Goddard test engineers believed the best way to wring out a spacecraft in the laboratory environment was to operate it through radio links, with no instrumentation connections, power, or other external connections that might cause or prevent a failure. JPL engineers, on the other hand, believed that they should try to exercise components individually and evaluate nonstandard conditions that might occur. This meant that there had to be synthesized inputs and special instrumentation connections to allow proper evaluation of subsystems or components.

One incident that occurred during the thick of competition for a Mars project involved Bill Stroud, one of the most outspoken Goddard engineers. He came to Headquarters one day predicting "doom and gloom" for JPL because of their approach, and offered to prove his point and help teach them how to do tests properly. For emphasis, he had with him a pair of gold-plated diagonal wire-cutting pliers, called "dykes" by technicians. According to Bill, all we had to do was take the dykes to JPL when they were running a systems test, cut the many wires and cables they were using to support their simulations, and then we would find out that their spacecraft would not work. The inference that JPL cheated on their systems tests, plus the "know-it-all" impression his act created, went over like a lead balloon with JPL engineers when they heard what Bill had done. I recall being more amused than concerned, for both centers had proven their competence, and I was sure that either approach could be made to work. Of course, this episode did nothing to encourage commonality in testing techniques among centers, and I still do not know which philosophy is best, if there is a best.

As is often the case with such conflicts in judgment, tradeoffs were made and compromises struck. In time, though, the balance moved toward testing everything that flew, subjecting it to as nearly complete a lifetime simulation as possible. This gave us confidence that the equipment was flight-ready, and I believe the principle was borne out by the successes that followed.

A particular bedevilment of those times arose from the requirement to sterilize everything that might land on the Moon or planets. No one objected to the idea of avoiding seeding them with terrestrial microorganisms; the problem arose from the fact that the specified protocols were complex, rigid, and at war with the quest for reliability. A sure and first specified approach was heat sterilization, but many components would not survive being baked in a hot oven for long periods. A less obvious complication of sterilization developed from the fact that it had to begin at the subassembly stage and be maintained through final assembly and testing, thereby requiring that all testing equipment and facilities be of clean-room quality, including handling gear. In an environment where even the hoists are sterilized, and kept so, work does not advance with the speed and precision one might want.

I always took the position that a spacecraft should be built at the factory, checked out there to the fullest extent possible, and shipped to the Cape ready for launch. Most engineers tended to agree with this philosophy; however, special circumstances always seemed to dictate the need for a complete systems checkout at the Cape, requiring equipment for a thorough test of all finally assembled and adjusted hardware just before mating to the launch vehicle. The field checkout facilities were identical in many ways to the system test facilities at JPL; in some respects they were more complex because of the need to include launch vehicle and tracking elements. The apparent duplication between these facilities was not in fact real; they often performed complementary tasks on the same entities for different purposes. Checkout could indicate a need for replacement, and with the window inexorably approaching, return to the factory might be unthinkable.

The first spacecraft assembly and checkout building used for Rangers and Mariners at the Cape was Hanger AE. It had been built in the 1950s for Navaho missile preparations and had a low-bay area designed for a flying vehicle resting on a tricycle gear. It was a non-air-conditioned, metal building with small shop areas alongside the hanger portion and was unsuited to the peculiar needs of spacecraft. Returning to the same facility that I had been associated with almost 10 years before was a ghostly experience. Engineers from JPL and the Cape quickly defined modifications, and I obtained approval for a high-bay area addition with a 30-foot hook height that would allow the spacecraft to be assembled and enshrouded vertically. Also included was an air-conditioning system with filters that provided the clean-room conditions needed for sterilization control.

When finished, it was one of the first clean facilities to be installed at the Cape, said to be cleaner by particulate count than most hospitals in the area. This appraisal led to its being named by Kurt Debus, Director of the Kennedy Space Center, "Hank Levy's hospital." Hank Levy was then, and still is, JPL's principal resident in charge at the Cape, a man whose fingerprints have probably been launched on more lunar and planetary spacecraft than anyone else's.

A significant amount of additional upgrading and modification went on, including changes to the spin-test facility, remote for safety reasons, to allow its use for fueling midcourse correction motors and to support the final ethylene oxide gas sterilization process. A new, designed-from-scratch systems test building was also begun. While prelaunch checkout facilities steadily became less ramshackle, it was 1964 before we could begin to treat our interplanetary travelers with the care they deserved.

At the same time, two other efforts of vital importance were being conducted. First, there was the establishment of a deep space network composed of radio tracking, telemetry, and command stations at different points around the globe, a control center from which it could be directed, and an Earth communications network to tie it together.

The Deep Space Instrumentation Facility (known universally as the DSIF) was a vital link in the chain. Obviously a good launch was not enough; mission success depended on good data return and analysis. The geographical position of the Earth stations, the communications frequencies to be used, the ground handling rates, and the priorities among spacecraft aloft were very real constraints that had to be factored into mission planning. The fact that two of the three tracking stations were on foreign soil, one of them subject to the vagaries of an unstable government, also led to occasional cases of heartburn.

Second, there was a flight operations facility at JPL, later known as the Space Flight Operations Facility (SFOF), with quarters and equipment for mission operations, including banks of computers for analyzing trajectories, acquiring and analyzing telemetry data, and generating commands to be sent to spacecraft. The SFOF necessarily had a very close relationship with the DSIF, and in fact they shared a common control center. In addition, there were launch operations facilities at Cape Canaveral for preflight testing of the systems for tracking and downrange support, plus all the diagnostic equipment needed to ensure that the launch phase was performed properly.

The man at JPL ultimately responsible for tracking, telemetry, and communications was Eberhardt Rechtin, a near-genius whose telecommunications achievements left his mark on space exploration. Rechtin had been a student of Bill Pickering in electrical engineering, graduating from CalTech with a cum laude doctorate in 1946. During the time the tracking and data acquisition network was taking form, Rechtin was a key figure, forceful and enthusiastic, with a reputation for brilliant solutions to technical problems. He had an uncommon knack for grasping the unanticipated implications of large systems, foreseeing both problems and potentialities ahead of others.

Rechtin was joined in the network's formative period by a number of other good men, among them Walter Victor, Henry Rector, William Samson, and Robertson Stevens; some of them are still making important contributions to the field today. The team envisioned three Earth stations located so that one of the three would always have a spacecraft in view as the globe turned. Thus, information could be received and commands sent without interruption. The stations are about 120° apart in longitude, one in the Mojave desert not far from JPL, one in South Africa (later replaced by a facility in Spain), and one in Australia near Woomera. Each site has large steerable antennas that can be pointed accurately in space and are designed for maximum efficiency for receiving and transmitting signals. At the beginning the preferred frequencies were from about 890 to 960 megahertz; signals from this region in the radio frequency spectrum pass through Earth's ionosphere without much reflection. Each station can transmit commands and receive data, in addition to establishing one- and two-way Doppler links for determining positions and trajectories of remote spacecraft.

The mission command post was the SFOF. First-time visitors found it a dramatic place, a large, essentially windowless building on a hillside, with a well-guarded entrance and a set of big diesel-electric generators down below. In a large, dimly lit room with multiple wall displays, the controllers on duty "worked" the distant spacecraft, while dancing numbers on the displays continually reported changing measurements. In adjoining rooms other engineers were concerned with their specialized areas, such as trajectory computation, data collection and reduction, and spacecraft engineering conditions. To visitors it was a paradoxical place: everything progressed inexorably and yet nothing seemed to happen; distances were unnaturally distorted, with Spain and Australia brought next door and an unimaginably distant spacecraft giving its speed and course with extraordinary precision. Even time was skewed: when the spacecraft reported an event, a visitor was bemused to realize that its "now" had occurred before his "now"; even at the speed of light, signals took several minutes to travel to and from distant space.

Of course the effect on visitors was the least of the concerns of those who designed the SFOF and obtained and programmed the computers that made it work. The man who deserves the most credit for this is Marshall Johnson, who came to JPL in 1957 as a computer engineer. Doing today what he managed to do some 20 years ago would be almost impossible, thanks not just to normal bureaucratic inertia but also to the encrustation of controls, loops, and reviews that sprout like rain-forest undergrowth. The computers had to be procured and installed; their software had to be written, checked, debugged, and specially adapted to the unique characteristics of the particular project; and then in a few months it all had to be done differently for another spacecraft on another mission. Marshall Johnson (and his staff) had the kind of wonderful competence that blossoms under tremendous pressure.

Transport of spacecraft hardware on Earth from JPL or from a West Coast factory to a Florida launch site was not simple; one does not simply nail a $50 million spacecraft in a crate and drop it off at the express office. Protection from contamination and from shock called for a controlled-atmosphere container traveling in a special air-suspension van on a route precharted to avoid low bridges and similar problems. Even then, there were the hazards of an occasional blowout, damage inflicted by irate snipers who didn't like "missiles," and the possibility of collision on the crowded freeways. It was something that schedule-minded managers learned by doing, worrying all the way.

The cost of disrupting human lives for unmanned spaceflights were far from negligible. There were questions of how personnel should be assigned to the assembly, checkout, and launch of almost-ready spacecraft and who should concentrate on developing the next one. The procedures continually evolved, but usually a large team had to spend the last six or more weeks before launch at the Cape. Families were split up or partially moved, with considerable hardship in either case.

Preparations for launch might begin with civilized 8-hour daytime duty periods, but as time shortened, working hours lengthened; there always seemed some critical milestone to be accomplished in the small hours of the night. To visitors, the preparations for launch seemed to go on in an informal but rather tense atmosphere. Foremen or supervisors were invisible among their subordinates—rolling up their sleeves, joining the workers, and doing what needed to be done. Many loved the excitement of the effort and were so caught up in it that they neglected their families; these were people who worked hard by preference and played hard for compensation. I always felt remorse that the toll on personal lives was so severe, but I knew no one who would have traded the experience of a successful space mission for any other.

A National Goal

Soon after becoming president, John F. Kennedy showed interest in the excitement of the space race with Russia and in the fact that the near-term inferiority of our launch vehicles seemed to condemn the United States to second place. He encouraged NASA leaders to focus on programs that might leapfrog serial developments and secure space preeminence for the country. Several ideas were studied, but the specific task of outlining a program for sending men to the Moon was assigned to a group of five NASA Headquarters members and two center representatives in early January 1961. Chaired by George Low, the group (made up of Eldon Hall, Alfred Mayo, E. O. Pearson, and myself from Headquarters, Maxime Faget from the Space Task Group at Langley, and Herman Koelle of the von Braun group at Marshall) reported after 4 weeks of concentrated study that a manned Moon landing was possible and could be accomplished under specified ground rules in 1968. Our brief study considered a number of options but recommended a direct ascent mission using a large rocket booster that was to be called Nova. The direct ascent concept called for a trajectory from Cape Canaveral to a landing on the Moon without either Earth or lunar orbit. The return from the Moon was to require a launch from the lunar surface directly to a reentry into Earth's atmosphere.

On the basis of the advice given him, President Kennedy made his famous speech proposing that the United States send men to the Moon and return them safely to Earth within the decade. The simple language and concise definition of a national goal was important and in itself a contribution to the final success of the Apollo program. To the public it offered the promise of a major space accomplishment in the foreseeable future, after a long string of past and probably future Russian triumphs. To the Congress it represented a clear goal they could discuss with their constituents and among themselves, if need be, when it came time to support it with funding commitments. To NASA and to the industrial and academic communities, it provided a focus on an activity that was soon to dominate the total space effort. Kennedy himself, as stated in a speech given at Rice University in September 1962, regarded it "...as among the most important decisions that will be made during my incumbency...." What an understatement!

Approval was not universal. To many who had begun to see the potential scientific rewards of space exploration and the practical uses of Earth orbit, the mandate for manned flight to the Moon was less attractive. An informal coalition of scientists decried the decision, arguing that far greater sums of money would be required, with lesser returns, than if funding were channeled into unmanned but exclusively practical and scientific missions.

Nor was some measure of disenchantment restricted to those who foresaw personal disadvantage; even within NASA, an agency with much to gain from the decision, there were elements that felt a manned lunar landing was a dubious goal. The widely respected Deputy Administrator, Hugh Dryden, had once noted in public testimony a parallel between suborbital manned flights and "shooting a man out of a cannon." T. Keith Glennan, the first Administrator of NASA and a man who had done much to organize and shape the new agency, beating off predatory forays by the military and establishing NASA's vaunted policy of complete openness about plans and results, was another with reservations about sending men to the Moon.

"It probably became apparent that I wasn't all that excited about man in space..." he told me in an interview 21 years later, "but it soon became apparent that we had to have the man-in-space program. To me the law said something—it said 'for benefit of all mankind.' I wasn't sure what man in space was going to do for all mankind very quickly."

But the times were changing and the tide was running. Dr. Glennan, an appointee of the previous administration, left NASA in January 1961. The public (and to some degree Congress as well) clearly reflected attitudes that were to prevail in subsequent decades: if the mission was manned, people cared deeply, and if only instruments flew, interest was lessened and somewhat remote. Even the most successful and rewarding planetary missions could never evoke the outpouring of fascinated concern elicited by astronauts.

Although the decision had been made and the goal set, unresolved questions about mission design remained. In the early days, many people at NASA (including the special task force I had served on) believed that direct ascent was the best approach to manned lunar missions. A group led by Wernher von Braun favored Earth-orbital rendezvous with launchings to the

Moon from, and return to, an orbital platform. Under the direct approach scheme using a very large rocket system, the spacecraft plus its landing and launching rockets was to be launched directly from Earth to a landing on the Moon, with a direct launch from the Moon directly to Earth atmosphere reentry.

Work was begun on the Nova rocket required to perform a direct ascent mission and continued until some time after a strongly worded letter was sent to Headquarters by a Langley research engineer named John Houbolt. He opposed both Earth-orbital rendezvous *and* the direct ascent scheme, arguing that rearranging a single vehicle in Earth orbit, launching into lunar orbit, descending to the lunar surface in a special vehicle, and returning to rendezvous in lunar orbit before the trip home represented a more logical plan. Though this type of mission appeared to be significantly more complex than a direct approach and return flight, it required a good deal less rocket energy, as it employed the effective concepts of staging to a maximum degree. Critics saw the scheme as "scattering hardware all the way to and from the Moon" but Houbolt's position was recognized as having a sound technical basis, and the issues were examined in more searching detail. After almost a year of analysis and debate, LOR (the acronym for lunar orbit rendezvous) was officially adopted.

Much initial work went into the development of efficient launch vehicles. Saturn launch vehicle hardware that could be built and tested in a stepwise manner was defined, leading to ultimate integration into a very large launch vehicle. Upper stages were visualized with some geometric relationships so that initial developments could be applied even though modifications were to be expected in the final configuration. Multiple engines allowed flexibility in design; we could combine as many as we needed for a particular stage. An early decision to develop a hydrogen-oxygen engine for upper-stage application was a significant technical choice.

With these concentrated efforts on high-performance rocket engines, the seeds were sown for later difficulties in the development of Centaur. Early work had been conducted under the auspices of the Air Force and ARPA (the Advanced Research Projects Agency) on high specific impulse rockets using hydrogen and oxygen. The NASA decision to develop this technology made it desirable for the work to be combined and assigned to Marshall Space Flight Center. This proved to be both good and bad for Centaur: good because it made sense to develop the hydrogen rockets for Centaur and Saturn under one roof, and bad because the keener preoccupation of Marshall personnel with Saturn meant that Centaur inevitably slipped into a second priority position. This later became a severe handicap for NASA's lunar and planetary programs. In reflecting on the attitude of von Braun and the Marshall team, Dr. Glennan described the situation thus: "Saturn was a dream, Centaur was a job."

Because of the Saturn-Centaur link, it may be well to review the status of Saturn at the time of the decision to make manned landing a national goal. Developments had been limited to Saturn C-1 and C-2 versions, capable of putting a small manned laboratory into Earth orbit. First-stage engines were to use existing technology, with liquid oxygen-jet propulsion fuel (a kind of kerosene) engines having less specific impulse—roughly half that of the high-performance hydrogen-oxygen engines planned for upper-stage development. During the time a direct approach to the Moon was contemplated, the huge new vehicle named Nova was also on the drawing board; it would cluster the large F-1 and J-2 engines under development, and some held that it might use large solid fuel rockets, then undeveloped. The Air Force had done some preliminary work on large solid fuel rockets; though they were far from ready to fly, it was contended that NASA already had its hands full with liquid fuel engine development and that the Air Force should continue large solid fuel rocket development. When the decision was made to accept Houbolt's LOR mission concept, it was possible to dispense with the gigantic Nova and all the additional complications its concurrent development would have brought.

Along with these decisions came some very significant budget increases for the lunar program managed by the Office of Space Science and Applications. Added funds were to strengthen unmanned exploration of the Moon, using hard-landing Rangers and soft-landing Surveyors to collect basic lunar information of value for the design of the coming manned landing spacecraft. Senior NASA officials considered this an essential preparatory step and took this position with Congress. Unfortunately, people in the Apollo program, dedicated to manned lunar landings, did not always agree, for a variety of reasons.

First, some had little confidence that unmanned spacecraft were capable of successful lunar missions. This attitude may have arisen in part from the self-confidence of a group intensely concentrating on a difficult time-limited goal, in part from a degree of pride not far from hubris, and in part from a cynical assessment of the string of failures of Ranger, the leadoff unmanned lunar effort. Some of the leading engineers on Apollo, including Max Faget (whom I had gotten to know during our special lunar study), took the position that the manned landing had to be planned without counting on *any* unmanned results.

Since the manned and unmanned programs were managed separately, with no common authority except at the administrator level, these differences in viewpoint were largely unnoticed. Those of us in Lunar and Planetary Programs did coordinate closely with systems engineers employed to support Apollo, mainly Bellcomm, Inc., experts from a division of American Telephone and Telegraph, who had been hired to conduct systems studies and to develop guidelines and tradeoffs for Apollo. In our meetings with Bellcomm some moderate conflicts arose occasionally, but rarely to a troubling degree.

Only once did the conflict detonate with a resounding report. The problem arose during a visit by Congressman Joseph Karth and others to the newly founded Manned Spacecraft Center in Houston in 1962. The Congressman, seated next to Max Faget at lunch, asked Max about Surveyor's importance to Apollo, in a context that implied that Surveyor had been funded largely on the basis of its probable importance to Apollo Lunar Module design. Faget, never known for pulling his punches, flatly told Karth they really were not depending on Surveyor. In fact, Max told him they had plans of their own for obtaining the necessary data by orbital reconnaissance with manned vehicles before committing to landing.

Decades later, after his retirement from NASA, Max described the incident in a reminiscing session we had in his Houston consulting office. Time had mellowed us both, so the story Max told about the incident did not seem as exasperating as it had originally. "I made a terrible mistake with Mr. Karth once," he recalled. "They were down here shortly after we arrived. Karth was, I realized afterwards, trying to justify some appropriations. We had an all-day-tell-them-about-the-program thing. Karth asked me, 'What kind of a problem would it amount to if the Surveyor program failed?' I said, 'That wouldn't be any bad problem. We can do it without those guys. We've got a great big wide landing gear and we just can't afford to be vulnerable to the loss of that program. We'd go ahead anyway.'

"I tried to explain to him the things we had. Within our own shop we had thoughts on what we'd do if we didn't get any support from the unmanned program. Actually, the unmanned program did several things. Ranger, of course, gave us a close-up view. It gave us some idea of the fine-grain roughness of the terrain, which was pretty important. And, of course,

Surveyor, by landing, proved that Tommy Gold [a Cornell University astronomer who had theorized that the Moon might have a surface of deep dust into which landers would sink and be lost] was all wet. I don't think that anybody really believed him. But we planned to make orbital flights if the other programs didn't come through—some *very* low orbits of the Moon. We had some penetrometers we were designing to drop from the spacecraft.

"We could make our own survey of the Moon, make our own penetrometers, and we were even talking about doing radar scans of the surface. In many ways it would have been a nice program to carry. We had a lunar survey module, a fairly large-diameter can that would replace the Lunar Module to allow us to spend as much as a week or so orbiting the Moon. It would have been a good program, but it didn't happen."

I reminded Max, "Well, I guess we heard about your conversation with Mr. Karth—it caused me to do a lot of writing and explaining."

"Oh, yes!" Max exclaimed, "Next morning Mr. Webb [NASA Administrator] called Dr. Gilruth [Director of the Manned Spacecraft Center] and gave him what for. Gilruth had to call me into his office. He was sympathetic but he said, 'I gotta tell you, Max, you really blew it.' He told me how exercised Webb was. Apparently Karth really gave Webb hell about it."

The conversation made waves for a time at NASA Headquarters. Webb promptly set the record straight about Surveyor's importance to Apollo and told Gilruth to make sure his people were properly informed of NASA policy in all external contacts thereafter. And so they were, to some degree. There were no more casual statements of independence, although I am not sure attitudes changed much. Those in charge at the Manned Spacecraft Center were still convinced that it was necessary to plan to obtain critically needed information with manned missions. They no longer spoke openly of doubts about Ranger and Surveyor, but they still held that their program could not depend on activities over which they had no control.

Apollo requirements were indeed high in the minds of those of us in the Lunar and Planetary Programs Office. We made every effort to ensure that the scientific mission objectives considered the urgent need for data to aid in the engineering design of Apollo. Obvious key questions concerned the nature of the lunar surface and its load-bearing strength. The resolution of the best Earth-based telescope photos at that time defined features the size of a football field—far too large for a confidently designed landing gear. Resolution on the order of 2 or 3 feet was a must. Early Ranger missions, in addition to providing TV coverage on approach that could give visual information, were also expected to eject a rocket-decelerated spacecraft containing a balsa-covered ball with a seismometer inside it. This was intended to survive impact and capture details about the structure and seismic activity on the Moon. The impact itself would allow inferences about surface strength.

Three Ranger flights of this type were planned, with the thought that at least one of the three could be expected to succeed. This kind of landing on a totally unknown surface was clearly risky, and some evolution on a trial-and-error basis was foreseen. We did not foresee that the early launch vehicles would not successfully deliver the Rangers to the Moon, and that when they did, Ranger spacecraft carrying landers would not work. It was not until July 29, 1964, that Ranger 7, the first fully successful flight in the series, sent back the pictures that justified that misfortune-dogged spacecraft.

Long after President Kennedy had established a manned lunar landing as a national goal, some measure of controversy lingered. To a few of the unconvinced, it was no more than a stunt—like going over Niagara Falls in a barrel or shooting gold bullion into space—certainly no basis for using tax funds. Fortunately, these jaundiced views did not prevail. Many foresaw that so broad and difficult an effort would inevitably create a great intellectual advance, filling gaps in knowledge of everything from algebra to zoology. Others saw it in terms of a national race with Russia, a competition for worldwide prestige in an area in which national dominance could be at stake. There were, however, those who argued that funds spent on Apollo could have been better spent right here on Earth, for schools and hospitals, dams and bridges. The Apollo missions, while not contributing to human welfare in the same way as a clinic or a highway, yielded significant advances in engineering methods and scientific knowledge. One of the peculiarities of the support of research is that, while specific, immediate benefit cannot be safely predicted, a multiplied social benefit almost always accrues. Except to those who argue from glib antithesis, knowledge is rarely evil; nor is ignorance a proper human goal.

Despite its critics, manned lunar landing was a steady and popular national goal. In a sense it exerted a unifying influence, almost the way an accepted war unifies the clamorous voices of peacetime. During the earlier part of the space race it seemed evident that the Russians were leading; this may have been a spur for us, in keeping with the observation that when you are number 2 you try harder. On the other hand, the American success with Apollo may have contributed to subsequent letdown and institutional dissolution, for high effort was no longer needed. Would it have been better for the national goal to have been more difficult and open-ended, for example, to explore the solar system and beyond? It does not seem likely that future goals as neatly constrained and defined will ever occur, but if they should, a greater degree of open-endedness could be desirable.

As the first unmanned lunar missions began, we were forced to come to grips with the thorny policy question concerning the degree of openness with which we would release the data acquired. From the beginning of NASA, Glennan and Dryden had been advocates of scientific openness, mindful of the language of the Space Act calling for ". . . the widest practicable and appropriate dissemination of information . . . " and ". . . for the benefit of all mankind. . . . " There was reason to believe that our new Administrator, James E. Webb, agreed significantly with his predecessors. However, when at last Ranger returned close-up photographs of the Moon and when Surveyor and Lunar Orbiter began to return a torrent of detailed new data, the strong military background of some people prompted them to argue the case for constraining the information. Lunar data might greatly aid the Russians, they argued, and in a race one does not present one's opponent with any assistance. Ingenious compromisers proposed intermediate positions of selective release and delayed publication, but the basic open position proved strongest, and all Ranger photos were promptly made available in atlases for the world's observatories, libraries, and technical information centers. It was, in retrospect, a wise decision, garnering respect and support worldwide for NASA and the United States.

Studied with care by the specialists at JPL and at the Manned Spacecraft Center, those first successful Ranger pictures in the summer of 1964 gave comforting information on the size and distribution of craters and rocks. They gave us confidence in the engineering model used for the design of landing gear. Some debate was still possible on the bearing strength of the surface, however, and it was only after Surveyor 1 soft-landed in 1966 that anxieties on this aspect were entirely set to rest. From the viewpoint of some onlookers, the *confirming* of assumptions about the Moon was less dramatic than their overturning would have been, but this is, of course, not the way engineers are trained to think.

As Surveyors continued to succeed—five of seven soft-landed on their lunar targets—personnel at the Manned Spacecraft Center were pleasantly surprised at the results. Among them was Max Faget, who had taken the position that he could not count on these unmanned spacecraft when designing the manned lunar vehicles. After Surveyor 1 soft-landed in working shape, Max called me at Headquarters to congratulate us and to say that he hadn't believed we could bring off an unmanned landing, especially not on the first try. Though we reveled in Max's "eating crow," we respected him greatly and took his words as high praise for our mission's work.

Three years later, those of us who had been involved in Surveyor felt special pleasure and a certain pride when Apollo 12 landed close to Surveyor 3, and its astronauts, Charles "Pete" Conrad, Jr., and Alan L. Bean, walked over to the silent spacecraft, took pictures of it with the Lunar Module in the background, and brought back parts to Earth for analysis. In the end, Surveyor's importance to the Apollo program could not be denied.

Close-up photos taken by Ranger's cameras dispelled many uncertainties about the size of boulders and craters. Lunar Orbiter missions were notable for mapping the surface and for helping to certify sites as suitable for manned landings. In addition to providing primary maps for all Apollo landing sites, by-product orbiter images, particularly the oblique photographs, allowed flight simulators to be developed that would help train astronauts to steer through the awesome terrain they would see as they descended to the surface. Lunar Orbiter also provided most of the knowledge we now have of the side of the Moon that never faces Earth. Surveyors gently touched down at five different sites (including the inside of a crater) to examine the strength, physical characteristics, and chemical constituents of surface material. They provided a wealth of information, later complemented by the soil and rock samples brought back by astronauts, that contributed much to our basic knowledge of the properties of the Moon.

The unmanned and manned lunar programs provided scientific data in a mutually reinforcing manner, with only modest overlap. All told, 13 successful unmanned and 6 manned spacecraft combined to produce most of our current knowledge of the Moon, assembled in a logical fashion that has withstood the test of time. Very few second guessers, if any, have shown ways in which the national goal might have been more efficiently achieved.

Ranger: Murphy's Law Spacecraft

If it is proper to feel sorry for spacecraft, Rangers deserved sympathy. Like the firstborn of pioneers, they had inexperienced and distracted parents with great expectations, and a wholly unknown environment to cope with. Worst of all, they had a perverse affinity for that malicious principle credited to Murphy: if anything can possibly go wrong, it will.

When the first definition of a lunar program came about, the United States was just beginning to organize the National Aeronautics and Space Administration into the singularly competent organization it would become. New people were being hired and sorted into teams, new managerial structures were being invented, and new standards of planning, quality control, and operations were being developed. At the nucleus of the new agency were groups of able engineers from three or four predecessor organizations who could draw on prior experiences of a somewhat related nature, but no one had ever done what NASA set out to accomplish. This time of ferment was not limited to the new agency. Widely spread in both industry and in academic communities were pockets of able, hard-driving people eager to find reputation and reward in the newly accessible territory of space. The military services were much a part of the scene, drawing activities from ballistic missile programs. Born in this turbulent period in 1958 and 1959, Ranger was still struggling for success 3 years later when its Mariner derivative succeeded in visiting the planet Venus.

The Ranger program was particularly bedeviled by the fact that its launch vehicles were being developed and perfected concurrently. Earlier Pioneers 1 to 4 (all lunar spacecraft that were launched into space but did not reach the Moon) had flown on Thor/Able and Juno intermediate-range ballistic missiles of very limited payload. Mission and payload design studies conducted by JPL in 1958 influenced NASA's decision to use an Atlas ICBM (rather than a Titan), since the Atlas flight test phase had already started and was well along. Unfortunately, an upper stage suitably matched to the Atlas for missions to the Moon and the planets did not exist. The initial plan called for JPL to develop a new upper stage to be called Vega; combined with Atlas it would be capable of launching Earth satellites and deep space payloads of 500 to 1000 pounds. Then, on December 11, 1959, NASA canceled Vega and directed JPL to relegate to research status its work on a 6000-pound-thrust Vega engine. A study group that included members from JPL led to the establishment of the Atlas/Agena B vehicle program as a replacement; the integration of this new vehicle combination was placed under the direction of the Marshall Space Flight Center.

The initial plan to use Vega had an impact on Ranger design, not simply because of the change from one vehicle to another but also because Vega and Ranger had been conceived as an integrated set. The Vega was to have had six longerons—the fore and aft framing members—and the Ranger spacecraft was designed with a matching hexagonal symmetry to ensure the lightest carry-through structure from the launch vehicle to the spacecraft.

When the Agena B replaced Vega as the upper stage of the launch vehicle, a considerable amount of work had already been performed to maximize payload weight and to relieve liftoff time constraints. To effectively launch a spacecraft to the Moon it was necessary to include a variable coast phase—sometimes called a "parking orbit"—and the Agena B's existing restart ability made this possible without additional staging. The existing guidance systems had sufficient precision, provided the spacecraft could make a midcourse maneuver for trajectory correction.

As specifications were finally established, a standard USAF Atlas/Agena B could, with minor modifications, carry a lunar spacecraft weighing 700 to 800 pounds. This vehicle was the only promising and, we thought, reasonably developed capability at hand. It was not ideal for interplanetary missions; the staging arrangement was unconventional, joining a 1½-stage vehicle with a restartable second stage. A three-stage launch vehicle would have been more suitable, but the time and costs that would have been required for development were prohibitive. Despite being a combination of an ICBM designed to deliver a 1500-pound warhead on a 5500-nautical-mile trajectory and an upper stage intended to supply orbital velocities after launch atop an intermediate-range Thor missile, the Atlas/Agena was, by the standards of the time, a formidable booster.

The Atlas, developed in the mid-1950s by General Dynamics, stood some 66 feet high, weighed 130 tons fueled, and had a sea level thrust of 370 000 pounds. The half stage consisted of two big Rocketdyne engines to be jettisoned 2 minutes after liftoff; the remaining first stage was a single large Rocketdyne sustainer engine, supplemented with vernier engines to fine tune the velocity. The bulk of the vehicle was taken up by giant propellant tanks containing liquid oxygen and RP-1, a kerosene-like fuel. So thin were the tank walls that the erected Atlas would crumple from its own weight if the tanks were not filled or pressurized. The Agena, developed in the late 1950s as part of an Air Force satellite project and modified for NASA space mission use, was powered by a Bell rocket engine of 16 000 pounds thrust. Its propellants were two unsavory chemicals known as unsymmetrical dimethyl hydrazine and inhibited red fuming nitric acid.

The transition of responsibilities for launch vehicles from military to space users began in the early days of NASA. High-level negotiations between NASA and Air Force officials initially focused on launch vehicle procurement and launch responsibilities. The fact that all potential space vehicles at the time were outgrowths of missiles meant that the military had been in control of these developments. The formation of NASA as a civilian agency gave it authority to develop vehicles, but there was no prudent way to begin without working out arrangements with the military for an orderly integration of requirements and procurements. The Air Force initially said regarding the Atlas/Agena, "Don't worry about a thing, NASA, we'll put you FOB on orbit," meaning that they would accept total responsibility for launch vehicle procurement, launch, and operation. This proposal was not accepted, and meetings continued at all levels until agreements were signed allowing NASA to have its own contracts for vehicle modifications and establishing NASA-controlled launch operations with military support. There was, however, a truly difficult transition period that lasted well into the mid-1960s.

The Atlas/Agena B launch vehicle and associated facilities, including launch-to-injection range support, were originally under the cognizance of the NASA Marshall Space Flight Center. This was an inherited responsibility, however, and Marshall procured the vehicles through the Air Force Space Systems Division (AFSSD). AFSSD administered the contracts and directed the contractors so that Marshall "could obtain maximum benefit from established Air Force procedures" and so that interference between the NASA programs and high-priority military programs was minimized. Lockheed Missiles and Space Company supplied the Agena stage and was the vehicle system contractor responsible for such areas as structural integration, trajectory and performance analysis, testing, operations planning, and

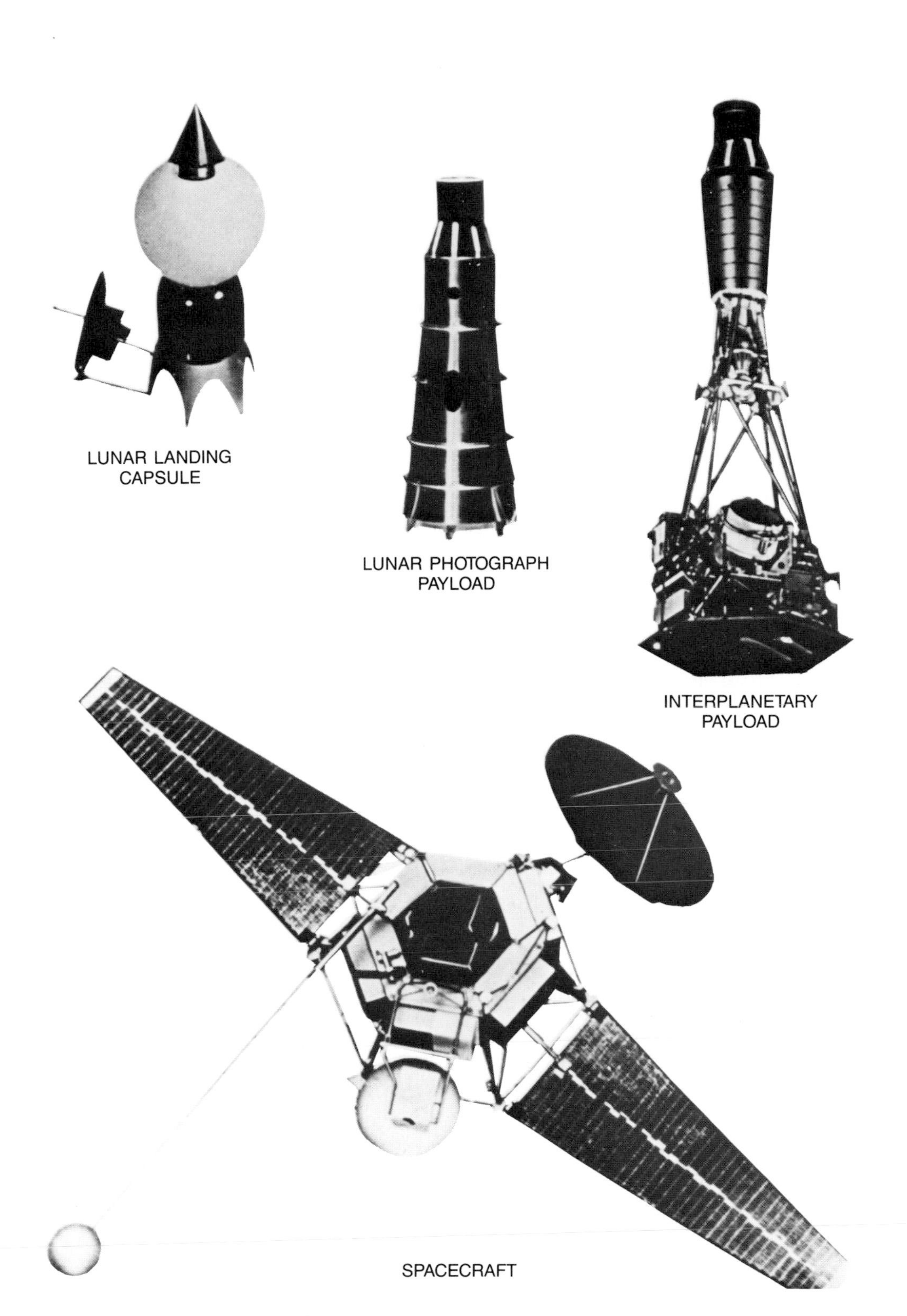

Ranger 3 mission payloads

documentation. General Dynamics/Astronautics was the contractor for the Atlas; both contractors had complete responsibility for their stages, with some uncertain assignments of responsibility for their integration. Marshall was supposed to integrate the two stages built by contractors, and JPL was to be responsible for overall integration of the vehicle with the spacecraft.

The Marshall assignment was a tough one; there were coordination problems and problems related to the inherited contractor arrangements, and the tasks were interwoven with Air Force activities. The Air Force actually had a more complex organization than NASA, with interfaces between its Space Systems Division and its contractors, and the Launch Operations Division and its contractors at Cape Canaveral. In addition, as already mentioned, within Marshall, the Atlas-based vehicles competed with the new Saturn program for personnel resources.

In August 1963, after work on the Centaur bogged down, responsibility for the Atlas-based vehicles, both Agena and Centaur, was transferred to the Lewis Research Center. Launch operations at the Atlantic Missile Range, previously under the direction of the launch operations directorate of Marshall, were reassigned to the Goddard Space Flight Center launch operations branch.

On December 21, 1959, NASA Headquarters sent JPL a detailed guideline letter outlining five lunar flights that emphasized obtaining information about the Moon's surface. The letter also requested that JPL "evaluate the possibility of useful data return from a survivable package incorporating . . . a lunar seismometer." This letter was the formal basis for what became the Ranger Program. Aside from the faint, ghostly Vega influences mentioned earlier, other influences shaped Ranger, some of them more useful to later interplanetary flights than to the immediate lunar mission. The Ranger concept called for a basic spacecraft to carry a variety of payloads, allowing development experience and costs to be amortized over several missions. Three different types of Rangers were developed. Although all were planned for launch on lunar trajectories, the first two were actually interplanetary spacecraft intended to obtain scientific information at great distances from Earth, with goals of developing the basic spacecraft technologies and adapting to the new Atlas/Agena launch vehicle.

The second block of three Rangers, more sophisticated in concept, were intended to make scientific measurements, including gamma-ray spectrometry, on the way to their destination, to take TV pictures on approach, and to land survivable capsules containing seismometer payloads on the

Moon. Though none succeeded, the attempt was brave and technically interesting, paving the way for future successes.

The last four Rangers had the specific goal of returning close-up pictures of the Moon's surface, using six television cameras capable of returning thousands of frames during the last 2 minutes before crashing on the Moon. While somewhat less ambitious technologically than landing survivable payloads, this block of Rangers produced three brilliant successes and represented a maturity in planning that comes with experience.

It should be noted that the block concept used with Ranger was derived from proven practices in the aircraft industry. Efficiency and reliability were believed to be best served by maintaining a constant configuration for a series or block of articles, only allowing new developments or modifications into grouped changes to the "production line." Although it was hard to see the merits of this concept from the early Ranger results, I still believe that it paid off, considering the total lineage of Ranger and Mariner hardware. In spite of this production-based philosophy for manufacture, I always tried to get the project team at JPL to regard each spacecraft as if it were the only one we had, vainly trying to achieve 100-percent success from the outset. After several failures, that management position was questioned by our critics, some sarcastically labeling Ranger a "shoot and hope" project. But as far as I am concerned, 100 percent success was a basic aspect of NASA philosophy from the outset.

While not all the concepts factored into Ranger design were essential for a trip to the Moon, they made it a kind of forerunner of interplanetary craft. It had become evident that opportunities for engineering development flight tests would be very limited for this class of spacecraft because the costs would be so high, yet mission criteria demanded sophisticated spacecraft designs. Thus, there was little hope of reaching a high probability of success for a single launch unless major parts of the spacecraft design remained the same from flight to flight to permit development experience. All hardware could not be new at every launch, in other words, and still provide a high probability of success. This view, together with recognition that only a fraction of total spacecraft weight could be decelerated by a retro rocket for a landing on the Moon, led to the bus-and-passenger concept for the Ranger/Mariner spacecraft that has been a hallmark of lunar and planetary missions.

A Ranger approaching the Moon from the distance of Earth on a minimum energy trajectory would normally impact at more than 4500 miles per hour. After arriving within a predetermined distance of the surface, a capsule-equipped Ranger was to be oriented to align its capsule rocket axis with the vertical descent velocity vector, and a marking radar was to call for spinup, separation from the bus, and firing of the rocket. After rocket burning to completion, the reduced approach velocity would allow a capsule to fall to the surface of the Moon at a survivable impact velocity. It was a novel capsule system, weighing a total of 300 pounds, made up of a small solid rocket and a mini-spacecraft enshrouded in balsa wood.

After studies by three contractors, Aeronutronics, a division of Ford Motor Company, had been chosen to design, build, test, and deliver three flight articles for a total of $3.6 million. Included were a special solid propellant retro motor, a radar altimeter to bounce signals from the lunar surface and trigger the retro at the proper instant, a crushable outer shell capable of withstanding impact on hard rock at up to 250 feet per second, and a spherical metal instrument package floated inside in a fluid to distribute and dampen impact loads. The flotation feature in some ways resembled the design of an egg, known to offer impact protection by reason of the fluid inside; it also provided for automatic erection and orientation of the package after landing. The instrument carried was a single-axis seismometer; also included were signal-conditioning electronics, a transmitter to report to Earth, and batteries to provide power for a 30-day lifetime on the Moon. The crushable outside structure was developed after an extended series of engineering tests of a variety of materials. As it turned out, the best impact absorbers were made of balsa wood, assembled around the capsule with the end-grain about 4 inches thick oriented in a radial direction.

This sophisticated little spacecraft-within-a-spacecraft system was carried toward the Moon on three occasions. On the first try a launch vehicle failure spoiled its chances; on the second and third trials troubles aboard the Ranger bus brought it to naught.

For a time we had hopes of follow-up missions using the capsule concept for landing other kinds of payloads on the Moon. One was a facsimile camera that, after landing, would poke its head through the balsa shell for a look around; it was a simple device with a nodding mirror that would do a line scan of the lunar landscape. The camera system and its capsule were developed satisfactorily and tested on Earth, but the program was canceled before this system got a chance to prove its worth.

Ranger bore the brunt of the difficult constraint imposed by sterilization. Any spacecraft likely to land on the surface of the Moon or a planet, by accident or by intent, had to be free of Earth's pervasive microbial population. This scientifically imposed international requirement placed a tremendous burden on Ranger engineering and development, and greatly multiplied costs. Of course, it also took its toll on the useful life and reliability of sterilized components and is believed to have had a seriously degrading effect on early spacecraft performance.

As an aside, the stringent requirements for heat sterilization may have, in the long run, resulted in the development of more reliable hardware, in much the same way that insecticides cause insects to develop an evolutionary resistance. But early Ranger spacecraft had to pay the price of meeting these stringent requirements. While there never was any positive proof that sterilization caused difficulties with electronic gear in space, there were many reasons to believe this was a factor in the high initial failure rates.

One incident from my earliest association with the Ranger project comes to mind as an illustration of project "growing pains." On first viewing the prototype spacecraft, I noted that the superstructure supports for the omni antenna were just four tubes attached to the bus and sloping upward to the smaller diameter base of the cylindrically shaped omni. Having been trained as an aeronautical engineer with heavy emphasis on light-weight structural design, I was immediately conscious of a difference between this structure and trusses common in aircraft structures.

"Where are the diagonal members to react against torsion," I asked. "Oh, there won't be any torsion loads," was the reply given. I knew well that almost any combination of compressive and lateral loads would result in torsion on a tapered structure, but being new and very conscious of the delicate relationship that had been created by the NASA "takeover" of JPL, I merely registered concern and went on to other matters.

In just a few weeks the vibration tests of the structure showed the need for diagonal bracing, as the torsional deflections were very severe. Diagonal members were quickly added, and I learned of a situation I was to encounter again many times. My assessment of the problem was that there were two contributing factors: (1) There was so much high technology associated with the conduct of a space mission that JPL project officials didn't spend time worrying about freshman-level design problems, (most young JPL engineers were trained in electronics and may have had little regard for civil engineering courses like Statics), and (2) the academic management style then operating at JPL gave independent responsibilities to many inexperienced people who were expected to function without supervision.

This was vastly different from the aircraft industry style of management I was used to, where a Chief Engineer and a highly structured organization made all key design decisions from the top down. Of course, after twelve years in the industry environment I thought JPL would benefit greatly from more discipline, and spent much effort during the years that followed trying to make this happen. But, as already stated, evolutionary changes in the JPL management style came slowly, and largely as a result of failures in early projects.

During the prelaunch phase for Ranger 1, meetings at the Cape were extremely confusing, as AFSSD representatives, launch operations representatives, Marshall and JPL personnel joined on their first mission of this series. In addition to the two principal launch vehicle contractors, General Dynamics and Lockheed, there were several other contractors responsible for guidance, tracking, and other services. Those early Ranger meetings were difficult scrimmages, part of the process of assembling, sorting out, and breaking in a new team. By the time Mariner 2 was launched, many of the pitfalls had been discovered, and the Mariner team had some insight into what was necessary; however, four Ranger launches had borne the brunt of the transitional jumble.

One of the key people involved in early operations at the Cape was Harris M. "Bud" Schurmeier. Bud was chief of the Systems Division at JPL, which had three major project functions: (1) systems analysis, including flight trajectory design, orbit determination, and the overall analyses required to establish midcourse and terminal maneuvers, (2) systems design and integration for the basic layout of the spacecraft and the entire supporting elements, and (3) spacecraft assembly, systems test, prelaunch checkout, launch, and flight operations. This single division contributed the "core group" of engineers who were involved directly in all missions, and it was therefore part of Schurmeier's responsibility to oversee a host of activities at the Cape that were essential to both Ranger and Mariner missions during the early days.

Bud and I recall early prelaunch readiness meetings at Cape Canaveral when a room full of people, including the various contractors, Air Force, NASA, and JPL personnel, convened for status reports. These early meetings were initiated without a disciplined agenda and with some uncertainty about who was in charge; they were presumably to allow each group to report to the others where they stood in their preparations for launch. Many of the people did not know each other, and at first there was no clear understanding of the role each group was playing. The spokesmen around the room would gleefully report progress until someone announced that his group was having a small problem. Upon digging into details, interactions with others would usually surface, and the meeting would end with some people not having to tell of problems, pending resolution of those that had come to light.

The hope that someone else would become the "fall guy" and allow more time to fix things would sometimes encourage bluffing, which only succeeded in hiding a difficulty if no one else knew of the problem. However, since many interface situations existed, there were lots of ways to be found out. Alert project managers got people "calibrated" and were better able to "sniff out" the true situations. A few of the participants were forced to disclose their problems after all the initial reports around the room—including theirs—had been favorable; it soon became obvious that it was much less embarrassing to be completely forthright from the beginning.

In reviewing some of his early impressions of activities, Schurmeier described his first project meeting at the Cape in this manner: "That was a real eye opener in a sense. The thing that depressed me was that so many different groups and organizations were all involved in various ways. It reminded me of a bunch of ants on a log floating downstream, each ant thinking he was steering. With the bewildering array of people involved and incomplete knowledge of all the facets of the operation there was a tendency on the part of the project manager to say, 'If everything is getting done and going the right way, leave well enough alone, and I don't care who thinks he's in charge.' " If things had really been getting done, this view would have been acceptable, but results soon proved the hoped-for success to be a fantasy. Bud's successful involvement in so many of the key project activities eventually led to his selection as Ranger Project Manager, when, after five failures, it was decided that a strengthening of the project team and a change of principal leadership was required.

In May 1961 I signed the review of qualification tests and approved shipment of the first Ranger to the Cape. A multitude of things had to be done to check out the spacecraft, the launch vehicle, and the launch facilities after the spacecraft arrived. The first Ranger launch window was from July 26 to August 2, with about a 45-minute window each day. In addition to scheduling launches to match the lunar cycle, there was a frenzied environment of launches about the time that required schedule coordination with range services and other project offices. We thought everything was ready on July 26, only to find to our dismay that the Range Safety Officer, the "bad guy" who was charged with blowing things up if the boosters went astray, did not have the trajectory information he needed. This caused a 1-day delay, to be followed by another delay the day after because an Atlas guidance system malfunction came to light. A third day was lost when a guidance program error was found in the input to the Cape computer.

These delays before the first countdown got underway were just the beginning. When the first count was within about 28 minutes of T-zero the entire blockhouse was plunged into darkness. The time was about 5:00 A.M., and it turned out that a short in primary power had been caused by the contraction of new power cables that had made contact with old conductors not yet removed. Those lines could be clearly seen, perhaps only a few hundred yards away from the blockhouse, when daylight came. Talk about frustrations!

Countdowns 2, 3, and 4 were scrubbed because of Ranger spacecraft checkout problems and another Atlas problem that surfaced. The final spacecraft failure—an electrical malfunction that triggered multiple commands from the CC&S—caused the spacecraft to be removed from the vehicle for repairs and the launch to be rescheduled for the next monthly opportunity in August. The only good thing about the frustrating experience was that we still had our hardware; at least it was not in the ocean. This "happy thought" was only slightly reassuring at the time.

Ranger 1 was finally launched in August 1961, a test flight not aimed at the Moon but intended to go to lunar distance and beyond. The Atlas/Agena was for the first time trying an Earth-escape type of mission, but failed to put the spacecraft on the highly elliptical trajectory being sought. Instead, Ranger was injected into a low Earth orbit and reentered the atmosphere after 7 days. Postflight analysis suggested that the problem was a switch circuit controlling propellant valves. Ranger seemed to have performed right, though short viewing times and movements into and out of Earth's shadow did not provide a meaningful test.

The next flight 3 months later had similar objectives and was a disgustingly comparable failure, with orbit at an even lower altitude and reentry after a few hours. This time the problem was overheating of some critical wiring in the Agena during the parking orbit period. After corrective action had convinced engineers that neither of these failures would recur, Ranger 3 was launched with considerable confidence, targeted to hard-land a capsule on the surface of the Moon. This time some booster circuitry that had behaved satisfactorily on the two previous flights failed, and the spacecraft was accelerated to a much higher velocity than desired, causing it to reach the Moon's orbit ahead of schedule and to miss the Moon by more than 22 000 miles. This prevented a true test of the hard-landing system, but when an attempt was made to exercise the landing system through commands, wiring problems were revealed that would have compromised the results even if the launch had been successful.

Finally, the launch vehicle for Ranger 4 performed beautifully, and launch computations revealed that the spacecraft would arrive at the Moon with no further course correction. Elation at this news was short-lived; engineering telemetry soon revealed that the spacecraft's central computer and sequencer, the heart of its control system, had lost its clock and could not perform the timing functions necessary for midcourse correction and approach maneuvers.

After all the earlier troubles, this mission looked somewhat better because of predictions that the spacecraft would impact the Moon. The Russians had sent a pennant to the Moon 2 years earlier, and Nikita Khrushchev had chided us publicly by quipping that their pennant had gotten lonesome waiting for an American companion. Administrator James Webb, who was in Los Angeles for a speaking engagement, was escorted to the Goldstone tracking station by Bill Pickering and me to witness the tracking to impact. This was a dubious honor; I would have enjoyed the trip more if we were to see a successful landing, but Mr. Webb was very gracious about supporting the team publicly in a press conference that followed.

We never will know the cause of the malfunction, but extensive engineering redesigns were made to the CC&S that prevented such a problem from recurring. Ranger 5, the last of the block of spacecraft intended to provide a survivable landing on the Moon, was launched in October 1962, shortly after Mariner 2 had begun its long trip to Venus. Ranger's launch vehicle performed well within the desired accuracy, placing the spacecraft on a flight path that would come within 450 miles of the Moon, easily within the capability of the course-correcting rocket onboard. But soon after the spacecraft was oriented so that the Sun would illuminate the solar panels, engineering telemetry reported a malfunction, probably in the switching circuitry for use of solar power.

After a few hours the batteries were depleted, and the spacecraft could not respond to commands to fire the rocket that would have placed it on a collision course with the Moon. By the time it reached the vicinity of the

Moon, all systems aboard were dead except for a small beacon in the landing capsule, poignantly reporting the position of the powerless craft as it sped past its target.

The turmoil generated by this uninterrupted series of failures was, at the least, considerable. In addition to blaming the launch vehicles for the problems they caused, there was considerable distress at JPL over problems with integrating scientific instruments that had little to do with the Moon into Rangers. At Headquarters we reluctantly agreed that trying to do too many things simultaneously was distracting, and it was decided that future Rangers would carry a payload of TV cameras that would concentrate on taking high-resolution pictures before impact. Since the position of the crash site would be controlled, and since the airless environment would not be conducive to propagation of biota, the sterilization of electronic parts would be relaxed. It was further decided that, even though it would delay the program for a year, a comprehensive review and redesign would be performed, along with a more intense testing program, all focused on maximizing the chances of success on the next attempt. In addition to these policy changes, Bud Schurmeier was asked to take charge as the new Project Manager.

Some 9 months later, just before Ranger 6 was to be shipped to the launch site, a company working at Cape Canaveral on a missile guidance system encountered failures in a type of diode that was also used extensively in Ranger circuits. The diodes, tiny units less than half an inch long, employed gold-plated elements encapsulated in glass. It was discovered that infinitesimal flakes of gold sometimes peeled off inside the capsules and floated around in zero gravity into positions that short-circuited the diodes. The culprit flakes were of microscopic size and generally made trouble only in zero gravity, but the suspect diodes had to be replaced. It took 3 additional months to replace and retest to make certain that the reworking had not inadvertently caused new problems.

In late January 1964, Ranger 6 and all its systems seemed ready. The launch appeared highly successful, the midcourse maneuver was executed precisely, and it was clear that the spacecraft would impact very close to its selected target site on the Moon. We eagerly awaited camera turn-on and warm-up, due some 15 minutes before impact. What we did not know in those heart-stopping moments was that the cameras *could not* be turned on, that during the first 2 minutes of launch the rocket had passed through clouds, picking up a charge of static electricity that had arced through the switch. Ranger 6 crashed close to its lunar target with its electronic eyes tightly shut. This failure was a bitter disappointment, the more so because success had seemed so near.

The considerable achievements of the launch vehicle and the spacecraft were almost totally eclipsed by the failure to return pictures, and Ranger critics rose up in numbers to disclaim the value of the effort. Detailed investigations far beyond ordinary failure reviews were instituted by NASA senior administrators and by congressional committees, and people working hard to keep a number of NASA programs moving were called on to explain all the problems from the beginning of Ranger to the present. It is a tribute to the Ranger team that they were able to cope with their own analyses and necessary rework while undergoing intensive management reviews and congressional investigations. For a time there was a question about whether Ranger would be terminated as a complete failure.

A now amusing incident occurred during a congressional review that was symbolic of the times. Bud Schurmeier and I spent two days before a Congressional Oversight Committee describing the spacecraft systems, tests that had been done, and other technical facts relevant to the Ranger 6 failure. During a discussion of the camera turn-on circuit that had apparently failed, I referred several times to the "common" switch that allowed the redundant channel to be activated in case the primary failed. Mr. Karth interrupted me to ask pointedly about our poor judgment that caused us to place a common, garden-variety component in such a sophisticated, multimillion dollar spacecraft.

After a moment of stunned silence, I realized that he had been misled by my use of the term common. The fact was, the switch was a high technology, solid-state device that was affected by the thousands of volts produced by a lightning discharge. It was typical Ranger irony that this necessary single element in an otherwise redundant system had failed; this simple miscommunication with the congressman made me realize how desperate we had become.

Any recounting of Ranger experiences would not be complete without some mention of Bill Cunningham and his involvement through thick and thin, from beginning to end. Bill, christened Newton William Cunningham, had joined Ed Cortright and the three or four others involved in lunar and planetary program activities at NASA Headquarters a few months before I did. Although he was hired mainly because of his scientific training in physics and meteorology, Bill developed managerial skills that helped bridge many a chasm while dealing with the tough challenges of Ranger. When I was named to direct lunar and planetary programs in 1961, Bill was named Ranger Program Manager. We had been working side by side; from that time on it was to be more like shoulder to shoulder.

Bill's dedication and loyalty were unmatched. Although I sometimes felt he was too forgiving and informal in his dealings with JPL, I knew that he provided qualities complementary to my more serious and sometimes tyrannical methods. On many occasions he was the "Indian scout" who restored peace and helped assuage the bitterness of JPL personnel who felt oppressed. When Bud Schurmeier became Ranger Project Manager, Bill worked closely with him to restore project relationships with NASA. They were very compatible, both on and off the job, and made my life much easier than it had been.

The three of us were together a lot during the 1960s, some of the time with our backs to the wall defending our project against hardware failures, against scientific critics, against adversary failure review boards, and sometimes against political committees and administrative fault-finders. Our association in countless meetings, at the Cape during launches, on airplanes and in airports, through the misery of six failures and finally the glow of success that came with Rangers 7, 8, and 9 cemented our friendship forever. I have often marveled at my good fortune in surviving the problems with Ranger and remaining as program director. My vulnerability, as "coach" of a losing team that was so much in the spotlight at the time, had made me continually aware of the debt I owed to colleagues like Bill and Bud and to my supportive superiors, particularly Ed Cortright. I kept trying to do the things I thought should be done, and by the grace of God, and with the help of these gifted friends, things worked out.

Ranger 7 was the first spacecraft to return close-up photographic coverage of the surface of the Moon. Aimed at an area chosen by scientific investigators as a candidate site for a manned landing, it provided the first sound evidence to validate the landing gear designs developed for Surveyor and Apollo. Ranger approached the Moon equipped with six TV cameras. A command turned on the two wide-angle cameras for warmup 18 minutes before impact, and four narrow-angle high-resolution cameras warmed up 15 minutes before impact. All cameras functioned perfectly, transmitting 4316 pictures before Ranger 7 crashed. The first picture showed an area of 500 000 square miles and the last, taken at very close range, showed an area 98 by 163 feet. The final images provided a resolution at least a thousand times better than the best pictures taken by Earth telescopes.

Ranger 8, also targeted toward a potential Apollo landing site, and Ranger 9, the last in the series and directed toward lunar highlands, were equally productive missions. Quite suddenly engineers concerned with the design of soft-landers yet to fly and scientists preoccupied with questions about the surface geology of the Moon found themselves almost drowning in a sea of new and superior images. Their varied reactions were perhaps predictable: the engineers found comfort in their existing landing gear designs, and the lunar scientists energetically demonstrated that the new data confirmed their preconceived, often conflicting theories about the Moon's origin.

The wisdom and confidence of those who decided to press on after six failures was borne out by the successes of Rangers 7, 8, and 9 and by the contributions they made to the lunar and planetary programs that followed. In fact, after the dismal failures, the tide decisively turned. The last three Rangers, followed by five of seven Surveyors (attempting vastly more challenging missions) and all five Lunar Orbiters, went on to perform their prescribed missions, and more, with outstanding success. From the perspective of time we can see that those six Ranger failures were not without reward; they taught us to organize and manage missions, to debug imperfect launch vehicles, to decide on and execute midcourse maneuvers, and to design, test, and launch spacecraft with a high probability of success.

Ranger also made incalculable contributions to what would shortly become useful new technologies. The diminutive rocket capsule designed to separate from Rangers 3, 4, and 5 and land on the Moon never got a chance, but the technologies evolved during its development were not wasted.

Although the project had a happy ending, Ranger sometimes reminded me of the ancient folk tale of Scottish King Bruce, repeatedly defeated by his enemies and on the verge of despair. While hiding in a barn, Bruce watched a spider trying to swing from one rafter to another, to spin his web from broad points of support. The spider tried and failed again and again; finally on the seventh try it succeeded in achieving its goal. This so inspired Bruce that he rallied his fugitive soldiers and at last won the victory that had so long eluded him. So it was with Ranger: six sickening failures before fortune smiled on the seventh attempt. For all its failures, Ranger paved the way for future lunar and planetary successes.

Launch of Mariner 2.

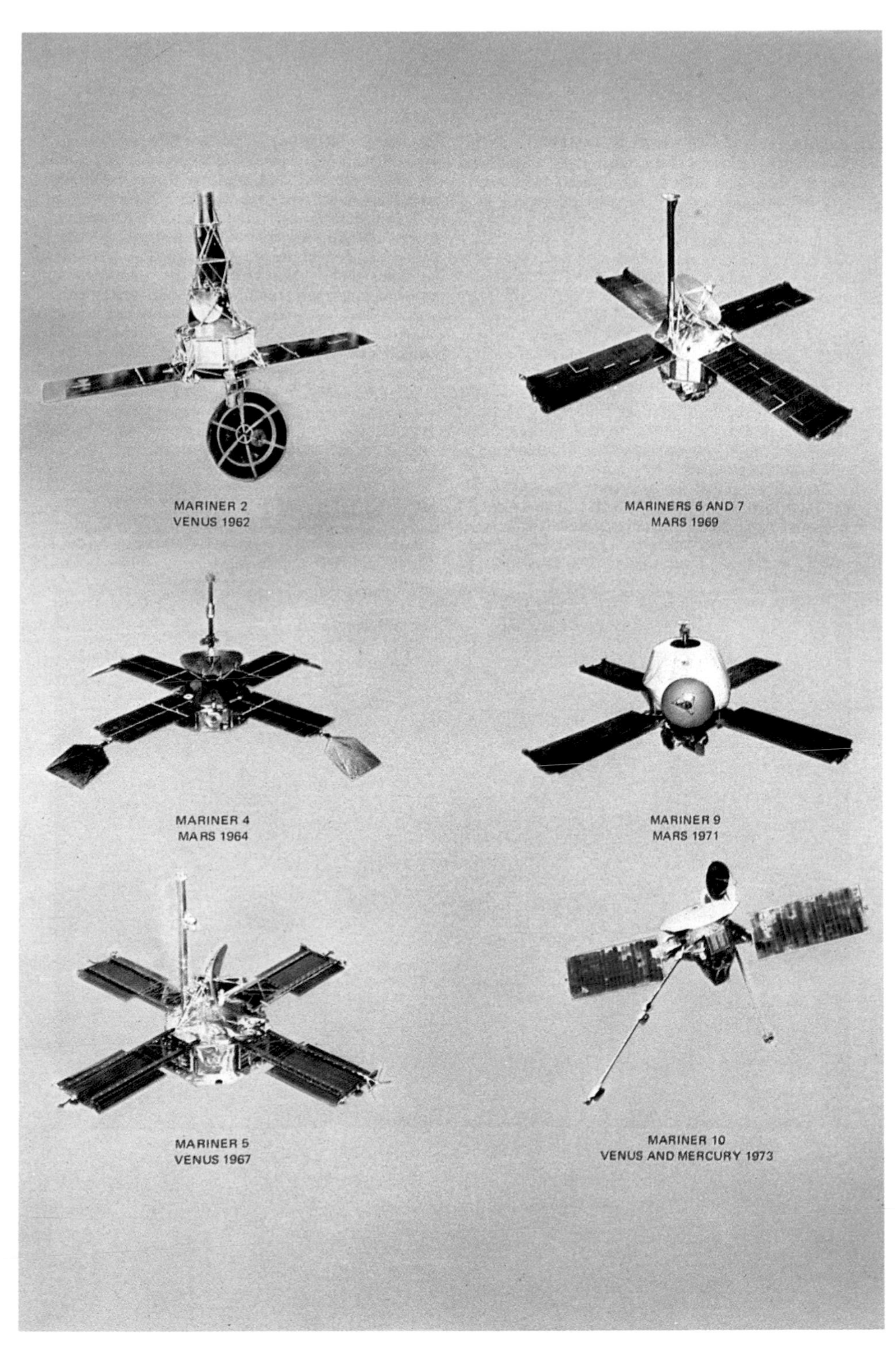

The Mariner spacecraft family.

Launch of Navaho missile in 1957. At the time, this 405 000 pound thrust rocket booster was the most powerful in the world.

Mariner 2 in the systems checkout facility in Hangar AE.

Astronaut Charles Conrad, Jr., examines Surveyor 3 on the Moon. The Apollo 12 Lunar Module landed about 600 feet from the unmanned spacecraft in the Ocean of Storms. Surveyor's TV camera and other instruments were returned to Earth by the astronauts.

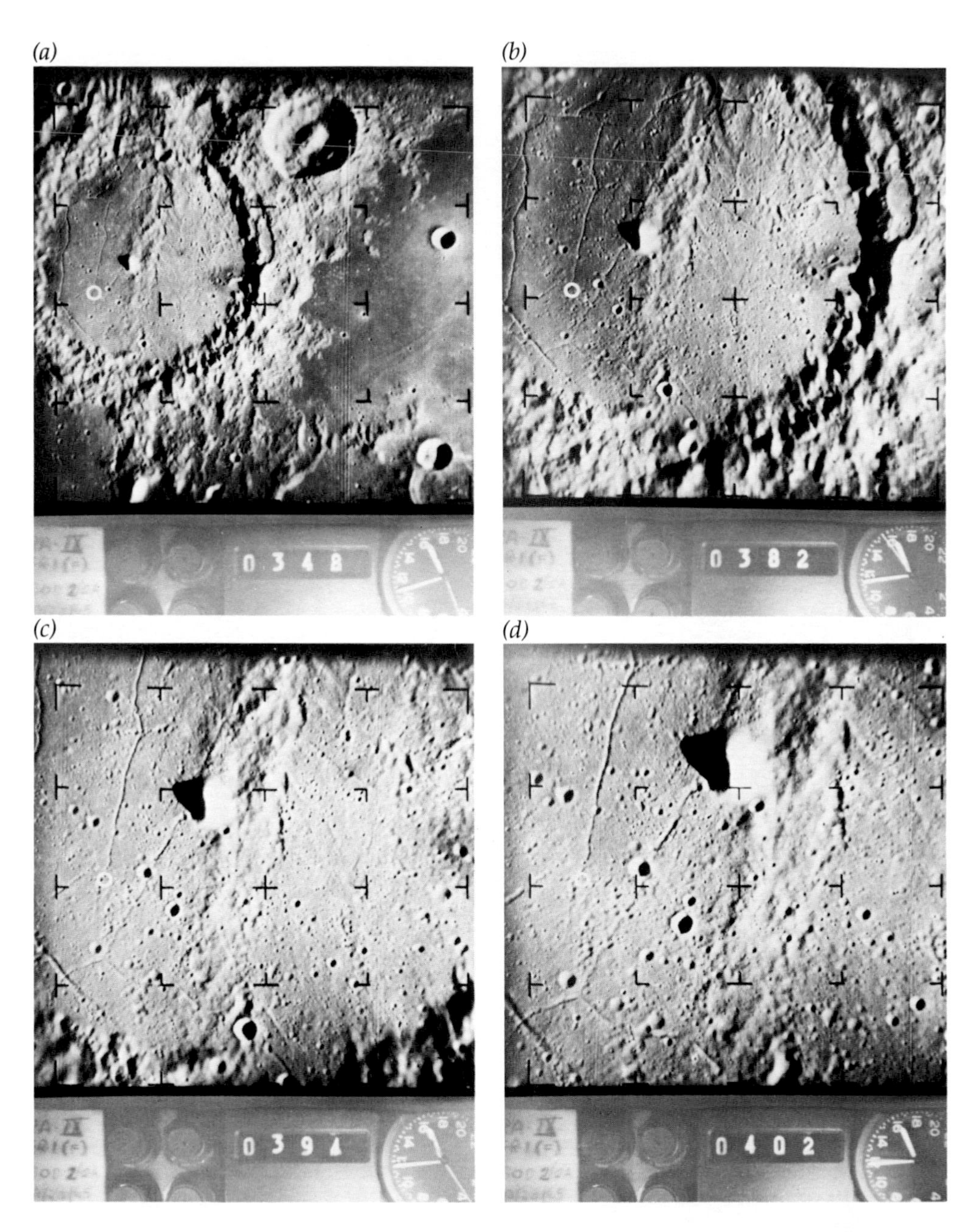

Photos taken by Ranger 9 as the spacecraft approached the Moon prior to impact. The white circle on each photo indicates the point of impact. (a) Altitude, 266 miles; time to impact, 3 minutes and 2 seconds. (b) Altitude, 141 miles; time to impact, 1 minute and 35 seconds. (c) Altitude, 95.5 miles; time to impact, 1 minute and 4 seconds. (d) Altitude, 65.4 miles; time to impact, 43.9 seconds; area shown, 31.6 by 28.5 miles.

Mosaic assembled to show the Surveyor 7 surface sampler digging a trench.

Surveyor drop test vehicle successfully lands on Earth.

The world's first view of Earth from the distance of the Moon, taken by Lunar Orbiter I during its sixteenth orbit.

First closeup photo of the crater Copernicus, one of the most prominent features on the face of the Moon. This oblique photo was taken by Lunar Orbiter's telephoto lens from 28.4 miles above the lunar surface and 150 miles due south of the center of the crater.

Workmen are dwarfed in the massive reflector of the 210-foot Deep Space Network antenna at the Goldstone facility.

This historic first closeup of Mars was made by hand at JPL as the picture was being radioed to Earth by Mariner 4. It was composed of 40 000 numbers (200 lines with 200 picture elements each) representing different values of grey, from white (0) to black (63). The picture numbers were printed sequentially on a strip of paper tape and then cut into picture lines. The lines of numbers were stapled, side by side, to a board, arbitrary colors were assigned to sets of numbers, and each number was colored with crayons by hand.

First color photo from the surface of Mars, taken by the Viking 1 lander. The horizon is about 1.8 miles from the camera.

View of Jupiter taken by Pioneer 10 from over a million miles away. The Great Red Spot and the shadow of the satellite Io can be seen.

Saturn as seen by Voyager 2 from 27 million miles away.

An astronaut in the manned maneuvering unit prepares to dock a satellite to be returned to Earth in the cargo bay of the space shuttle Discovery.

Essentials for Surveyor

Looking back with the help of detailed program reports and 20 years of additional experience, I view the goals, objectives, and achievements of the Surveyor program with a clearer perspective than when I was involved in its planning. Reliving Surveyor challenges and results gives me a warm feeling, for in retrospect, our team did not appreciate the engineering obstacles that would be encountered and overcome. Considering the scope of the total Surveyor effort and the new technologies required, we were very fortunate that five of the seven Surveyor spacecraft performed brilliantly.

Reports show that we did recognize spacecraft design challenges in a general way. However, in some planning documents the fact that Surveyor was to be launched with a newly developed Atlas/Centaur launch vehicle instead of an extensively tested and proven Atlas/Agena was merely mentioned in passing. Considering the problems that changes in specified performance for the new vehicle caused scientists, whose experiments had to be jettisoned because of the reduced payload, the scant reference to this problem now seems strange. At the time, the experience was almost as traumatic as if we were selecting by lot and throwing some passengers overboard at sea to save the sinking ship. The reductions in weight capacity of the Atlas/Centaur caused compromising modifications throughout Surveyor's development, making it abundantly clear that, for all its promise of greater performance, the new hydrogen-oxygen technology barely arrived in time.

The transit portion of the Surveyor mission from launch to the Moon was similar to the cruise mode for Ranger and Mariner; by 1964 there was some confidence in our ability to perform that phase. Nevertheless, the matter of achieving a cruise mode with orientation to provide solar power and ensure a midcourse correction maneuver—another rocket firing that required all aspects of attitude orientation and its many complexities—was never to be taken lightly. Not only had Surveyor to navigate through space between Earth and the Moon, but the landing on the Moon had to be made using a brand-new retro rocket, terminal radar systems, and modulated vernier rockets to allow the vehicle to soft-land 240 000 miles away from the humans who had designed it. The frightening part was that no one could make adjustments or do the little things that are often required to make rocket launches successful.

When I joined NASA Headquarters in 1960, Benjamin Milwitzky was already there. Ben had been a long-time NACA researcher at Langley before NASA was formed, specializing in structural dynamics and other sophisticated engineering activities. He was noted for his sharp technical skills and his meticulous attention to detail. He and I were assigned to work together on lunar flight systems and became close associates throughout the Ranger, Orbiter, and Surveyor programs. At the beginning of the Surveyor formal definition phase, Ben was named Surveyor Program Manager.

Ben's technical background in dynamics had included involvement in the design and analysis of aircraft landing gear, an area of major importance in developing a Surveyor that would land by dropping onto the unknown surface of the Moon. In addition to the direct applicability of his background to this and related engineering challenges, Ben had been well schooled in solving tough technical problems of any type. There were times when I felt his "research" approach to solving problems was at odds with our development tasks and management assignments, but hindsight clearly shows the tremendous benefits his talents, skills, and dedication brought to this undertaking.

The coordinated development of a Surveyor engineering definition was one of Ben's first tasks. Working closely with JPL, Ben and the project office established requirements for both mission and spacecraft that would be specified for bidders hoping to develop and build the spacecraft hardware. Because of JPL's commitment to Ranger and Mariner programs, both largely being built and tested in-house, we decided at the outset that Surveyor needed the participation of a prime contractor. This was not looked on with favor by most JPL officials; perhaps their laboratory background plus some frustrating experiences with contractors providing missile hardware were the reasons for their concern. Whatever the cause, they had reluctantly gone along with the strong NASA Headquarters position that a contractor would be used to develop, build, test, and support the Surveyor spacecraft.

After the project bogged down in midstream, the extent of Surveyor planning was criticized by congressional subcommittees and others. But considering how the spacecraft came out I believe it remained remarkably close to the concept envisioned during the formative stages. Actually, four concept proposals were presented by prime aerospace contractors; the one offered by Hughes Aircraft was chosen. It was primarily the technical aspects of the Hughes proposal that resulted in its selection, for all of us, especially Milwitzky and his JPL project counterparts, were technically oriented. Much later in the program we came to appreciate the importance of other factors, as we learned several management lessons the hard way.

It is difficult to put the necessary engineering tasks and technologies in proper perspective, but the new and most significant challenge for Surveyor was the landing. The three major elements of the landing system were (1) a high-performance solid rocket motor to provide the bulk of the velocity reduction on approach, (2) liquid propellant vernier engines capable not only of varying thrust but also of swivelling to allow attitude orientation, and (3) the landing radar system, which sensed distances from vertical and lateral motions with respect to the Moon. Because the allowable spacecraft weight for Centaur was only about 2150 to 2500 pounds, the retro rocket that decelerated the spacecraft near the Moon had to be very efficient. Even with a highly efficient rocket, the landed weight of the spacecraft would only be about 650 pounds, barely enough to incorporate the power, environmental control, communications, and scientific instruments necessary to make the mission useful.

The solid propellant retro rocket designed by the Thiokol Chemical Corporation and later designated the TE-364 was chosen because, in concept, it provided high reliability and simplicity. While simple in the operational sense, solid rocket design is far from a simple matter because the margins for error are so small. For the Surveyor retro, the case had to be as light as possible, or, to put it another way, the ratio of propellant weight to total weight had to be as large as possible. Efficient cylindrical cases had been made from spiral wound fiber glass, but for Surveyor the case was made spherical because it is the most efficient shape for a pressure vessel. Fiber glass was not suitable for the spherical shape, and steel was used.

The large expansion ratio nozzle was embedded as far as possible inside the case to shorten the rocket and to save weight. This involved some experimental development, but with good design and testing the Surveyor rocket produced the highest performance ever for such a large solid rocket. It was designed to produce a vacuum thrust of 8000 to 10 000 pounds with propellant loading to suit the final spacecraft weight and landing requirements. Its burn time was approximately 40 seconds, and it produced a specific impulse of about 275 to 280 seconds.

The three small vernier engines were also specially developed for Surveyor application by the Reaction Motors Division of Thiokol Chemical. These engines used hypergolic liquid propellants with a fuel of monomethyl hydrazine hydrate and an oxidizer of MONO-10 (90 percent N_2O_4 and 10 percent NO). Each of the three throttleable thrust chambers could produce between 30 and 104 pounds of thrust on command. One engine was swivelled to provide roll control to the spacecraft. The development of throttleable liquid rockets had always been a challenge because it was difficult to maintain proper fuel-oxidizer ratios and achieve reasonable performance with fixed-geometry thrust chamber and nozzle. Although the engines were small, obtaining repeatable performance and accurate adjustment was a significant engineering development.

To properly control the rocket systems, a radar altitude Doppler velocity sensing (RADVS) system was developed by the Ryan Aeronautical Company. This included a so-called marking radar, which initiated the signal to fire the main retro, and the closed-loop system, which provided signals for the operation of the vernier engines during the soft-landing. At an altitude of about 59 miles above the Moon's surface, a signal was generated by the altitude marking radar mounted within the nozzle of the main retro rocket, and 7 seconds later, at about 47 miles, ignition took place, expelling the radar unit and initiating the 42-second main retro rocket burn. At the completion of this burn and after jettison of the empty rocket case, Surveyor was close enough to the Moon to receive an excellent radar return from the surface. Operating in a closed-loop mode, RADVS sensors provided signals that were processed by the onboard computer and fed into the autopilot that controlled the three vernier rocket engines for steering and decelerating the spacecraft along a predetermined, optimum descent profile. Finally, at an altitude of about 14 feet, the vernier engines were cut off, allowing Surveyor to drop gently to the surface of the Moon, touching down at a speed of approximately 7 miles per hour. While providing the throttling of the engines to reduce the descent velocity, the radar signals also provided the information necessary to orient the spacecraft vertically and to diminish any sidewise motion relative to the surface of the Moon which might have caused a tipover on touchdown.

Sometimes we are given the impression that very large rockets like the Saturn are more difficult to design and build than small rocket systems like those employed in Surveyor. From an engineering and technology standpoint, this is not necessarily true. Indeed, the vernier retro system of

Surveyor and its closed-loop guidance using surface-sensing radar involved many technical facets that were more demanding than those required for control systems on a large booster rocket. In addition to the sophisticated technologies, the problem of weight constraints and size limitations produced additional challenges.

A further design concern for the Surveyor landing system that tended to be forgotten after the successful landings was our uncertainty about the surface of the Moon and its suitability for a landing. At the time Surveyor was being designed, theories about the composition of the lunar surface varied widely. The spectrum of opinions ranged from many feet of soft dust which would not have supported normal landing gear to large boulders and craters in such array that Surveyor could not have touched down without impaling itself or overturning.

The engineering model of the lunar surface actually used for Surveyor design was developed after study of all the theories and information available. Fortunately, this model was prepared by engineers who were not emotionally involved in the generation of scientific theories, and the resulting landing system requirements were remarkably accurate. Even with high praise for the generation of a realistic lunar surface model, however, I would be the last to say that Surveyor landings did not involve a certain amount of good luck. Indeed, photos taken at every landing site showed features within view of the cameras that could have caused catastrophic results if a landing had been made a short distance from the actual point of touchdown.

Without question, the approach and landing radar, the high-performance retro rocket system, the attitude control system, and the variable-thrust rockets required for landing involved extremely challenging engineering tasks that took somewhat longer and were more costly than initially envisioned. In retrospect, the actual development times and costs do not appear excessive, but in the 1960s, when so little was known of the entire process, estimates for the scope of the effort were far lower than they should have been.

One of the management decisions made in the development of the approach and landing system for Surveyor was to conduct simulated landing experiments on the surface of Earth with a system as nearly complete as possible. I advocated this plan, because I believed that we would seem foolish if problems occurred during landing on the surface of the Moon that might have been discovered during a simulated approach and landing on the surface of Earth. Of course, such a simulation involved tradeoffs: the gravities of Earth and the Moon differ by a ratio of 6 to 1, the atmosphere on Earth produces aerodynamic effects for a test descent that are not present on the Moon, and rocket performance in the atmosphere is different from that in a vacuum. It was thus necessary to introduce known compromises into the engineering of the model for Earth landings. Two aspects of the drop tests which proved to be extremely valuable were (1) the requirement for exercise of every landing system component in concert and (2) the necessity for subsystem team members to work out problems together through a realistic scrimmage before the actual mission on the Moon.

In general, the plan involved simple logic: a number of tethered tests would be performed, first using a large crane and later balloons, which would allow performance testing of the radar and spacecraft controls above the surface of Earth without the danger of a crash. The final test phase would involve 1500-foot drops from a balloon in which the Surveyor landing article would actually conduct its own descent phase, including landing on the surface. Three consecutive successful landings were declared to be mandatory to meet the goals of the test. The drop tests were conducted at White Sands, New Mexico.

Early landing system tests were not successful. One of the mistakes made initially and recognized later was that the hardware used for drop testing was not of flight quality in every respect. This was frustrating and time consuming because the test hardware that failed might not have been used in the actual mission and might not have failed. In addition to hardware shortfalls, the first tests were not conducted with the discipline and rigor that would have been present had the test landings been taken more seriously. After a significant amount of difficulty, discipline was introduced through special project-like assignments. People were told in no uncertain terms that they were to conduct the test activity as if it were a real mission. Incentive awards and other means of recognizing the importance of the tests were included in the plan. The final results were good, culminating in the required number of successful drops and providing as much proof as possible that the entire attitude control rocket radar landing system had been integrated well enough to achieve landings on the Moon.

In October 1965, just a few months before the first successful Surveyor landing in May 1966, a critical review of the Surveyor project was conducted by a House Committee on Oversight. Their report expressed concerns in closing paragraphs:

> Surveyor has undergone a great number of substantial changes. It can be expected of course that all complex research and development projects will undergo a certain number of significant changes as the work proceeds. The committee recognizes that such modifications are necessary to success and when executed in a timely fashion can contribute to costs and schedule objectives.
>
> The above documentary history, however, indicates that the Surveyor project has experienced an excessive number of extraordinary and fundamental modifications; the inevitable result of a poorly defined project.

While one cannot take issue with the generalities expressed in the document, I now feel that the committee's bold expectations for the project, probably encouraged by the confidence evident in our early planning documents, were perhaps inappropriate.

Seven years passed from project initiation to the final flight of the seventh Surveyor. In 1964, at the midpoint in development, technical and management problems were obvious. This was the year that the vernier engines contractor encountered such severe technical difficulties that the JPL project office terminated the contract with Reaction Motors Division (RMD), and sought an alternative source. This step was taken and the results presented to us at Headquarters as a *fait accompli* after acceptable progress seemed hopeless. Although upset by this precipitous action, we went along with the initiation of a new development contract with Space Technology Laboratories (STL) for replacement verniers. The gravity of the situation caused me to become personally involved, and one of the first things I did was visit both STL and RMD. It became obvious that we really were in a bind: the RMD hardware was in short supply, test results had been spotty, their manufacturing and test facilities were run down and poorly equipped (I remember describing the place as a "bucket shop" to my associates), and STL obviously needed time that we did not have to come up to speed.

As is often the case, however, the darkness was worst just before the dawn; at the time of the termination of the contract with Reaction Motors, it was true that a lot had happened without any indication of success for the rocket engines. Cancellation of the contract distressed RMD management, of course, and they were doggedly determined to carry the development efforts a step further. Their significant progress in turning the situation around (using their own funds, I might add) plus their willingness to reenter contract status on a negotiated basis were commendable. I believe RMD engineers and management officials made a remarkable recovery because of a genuine interest in the Surveyor project and because of a genuine concern for their company's integrity. In the terse language of the House Oversight Report:

> In any case, the history of the vernier engine development is noteworthy for two reasons. To begin with, a remarkable sequence of events took place in rapid order. First, JPL ordered the RMD work to be terminated and STL was placed under contract; then the laboratory reinstated the RMD contract and cancelled the STL contract; all within a period of less than four months. It seems fair to assume that this is an expensive way to do business.
>
> On the other hand, termination of the RMD contract seems to have had a salutary effect. Evidently, technical and management problems were solved in rather short order when the contractor realized what was at stake and that the government was willing to cancel his contract.

In the same year, the radar altimeter and Doppler velocity system under development at Ryan Aeronautical was experiencing severe technical problems. This system was definitely pushing the state of the art—it had to provide triggering for the main retro from an altitude of 50 to 60 miles and then provide the control signals for rocket orientation and thrust levels from approach to touchdown. All these functions had to culminate in a final sink rate of about 5 to 10 feet per second at touchdown! Looking back, it seems obvious that if such technology had already been available, helicopters would have been using it to land under poor visibility conditions. It is interesting to note that the techniques developed for Surveyor have not yet been incorporated into everyday use by helicopters, even after 20 years.

Also in 1964, two of the initial drop tests—in which Surveyor test vehicles suspended from a balloon 1500 feet above the surface of the desert were dropped to Earth—failed. In the first case, an electrostatic discharge apparently caused a failure in the release mechanism; the test vehicle was dropped prematurely and crashed. Thus, the failure was determined to be associated with the test environment only. In the second drop test in October 1964, five independent component failures were identified; some involved the spacecraft, and some were associated with test equipment. These failures prompted the effort to regroup and introduce discipline into the tests by using flight-quality hardware and better procedures.

In the first half of 1964, a nightmare period for Surveyor, other frustrations occurred. On January 30, Ranger 6 failed to operate after being launched successfully, triggering the failure review essential to recovery planning and initiation of engineering changes before the next Ranger flight. In addition, a congressional oversight committee held hearings in April on the Ranger failures. Preparing for and participating in that 4-day "inquisition" took a lot of my time and the time of several key people at JPL.

Selection of a contractor for the Lunar Orbiter had been made in December 1963, and we were inaugurating a new project organization at Langley in addition to negotiating a contract with the Boeing Company, new to the manufacture of lunar spacecraft. Because of congressional questions about the selection of Boeing, I had several "response activities" to deal with in addition to the contract preplanning and negotiations that were going on from January through March toward a new incentive-type contract. During that period it was necessary for me to travel to Seattle for conferences with Boeing and Air Force representatives and also to meet at Langley, Lewis, and JPL to encourage good field center management arrangements for the Lunar Orbiter. In January 1964 the Mariner Mars '64 spacecraft design was frozen, and NASA quarterly reviews were held in February and May as part of our management discipline. The Mariners were to be shipped in the summer for launch in November, and close coordination was required to ensure that test results met preshipping requirements. Surveyor alone presented plenty of problems, but I really had my hands full with failure reviews, contract difficulties, overruns, and the development of plans for additional projects.

Our own Headquarters project review of Surveyor 1 initiated in March 1964 produced a number of disturbing findings and recommendations. None of these really surprised me, but the formal returns from this review added more weight to our recommendations for action. Milwitzky, Cortwright and I had been advocating for some time the strengthening of the Surveyor project activities at JPL. Because of JPL's diverse in-house project involvements, we also felt that a deputy director or general manager who had more experience with contracting and related management matters was needed to augment the director's staff. Finally, after pressure was applied for several months, the CalTech Board of Directors encouraged Bill Pickering to hire retired Major General Alvin Luedecke as Deputy Director. Luedecke had been manager of the Atomic Energy Commission for several years and was planning to leave.

Luedecke's arrival at JPL on August 1, 1964, was welcomed by those of us at NASA Headquarters, and I immediately began to work with him on what I termed "recovery planning" for Surveyor. By this time, the success of Ranger 7 had improved NASA–JPL relationships somewhat, and General Luedecke rolled up his sleeves and addressed the Surveyor question as a major effort. Among the first things that occurred was the upgrading of the project staff, beginning with the assignment of Robert J. Parks, then responsible for JPL's planetary projects, as the Surveyor Project Manager. Eugene Giberson, who had been the Surveyor manager, stepped down but remained a valuable member of the Surveyor team. To his credit, he recovered from the experience with knowledge and skills that were later applied in the successful management of other major projects. Additional JPL staff members were immediately assigned to the effort because Parks already had a number of systems engineers and others under his aegis; with the decision to upgrade the Surveyor team significantly, some 200 people were assigned in short order.

It is interesting to note how good men rally to worthwhile causes in times of need. Alvin R. Luedecke appeared at a very propitious moment in the history of the lunar and planetary programs. As already mentioned, he was hired with considerable "encouragement" by NASA after the need for stronger discipline in making organization assignments and dealing with contractual matters at JPL was recognized. It would be easy for a casual observer to assume that General Luedecke could have made only a minor contribution during his few years at JPL; in my view, what he did was a keystone effort that resulted in significant long-term benefits.

For one thing, during 3 years at JPL, Luedecke's many 16-hour days and 7-day weeks amounted to 6 or 7 years of effort on a normal work schedule. He was on the job in the office much of the time, but he was never away from the work, as it was his nature to spend as much time as necessary on his tasks. After getting to know this impressive man personally, I learned that he brought to JPL many years of experience in tackling tough jobs and wrestling them to the ground.

A can-do attitude was evident from his early choice of a college curriculum to his last appointment as an acting university president. When he decided in 1928 to leave the ranch and attend college, he chose to study chemical engineering, partly because he was told it was the toughest and most challenging branch of engineering available at the time. Over the years his jobs seemed to lead him into the newest and least-known regions of technology because of the same drive.

An Army Air Corps pilot officer for many years, Luedecke became a general during World War II. When the war ended, he was directly involved in nuclear energy, the very newest technology at the time. His assignments included the development of weapons, facilities, and ranges for testing, eventually leading to his selection as general manager of the Atomic Energy Commission. From what I have determined, trying to manage that large, highly technical agency with its many critics, plus the problems of dealing with secrecy and intrigue, would have been a difficult chore in its own right. But I *never* understood how a general manager could survive in that environment while also reporting to a number of commissioners who were political appointees from all walks of life. General Luedecke managed to do this successfully for 6 years.

The change from nuclear energy to space was just the sort of challenge Luedecke liked to tackle, and I believe his experience, determination, and soft-spoken manner were just what JPL needed at the time. Of course his coming was not welcomed by most of the staff because of uncertainties about what might happen, but in time things settled down, as Luedecke's personalized, get-involved methods soon rallied support of his leadership. Bill Pickering took an extended trip shortly after Luedecke arrived; this gave Luedecke time to become acquainted with activities and key personnel, and precluded divisive game playing by disgruntled employees.

At NASA Headquarters it was recognized that a major upgrading of the Surveyor contract with the Hughes Aircraft Company was required. Ed Cortright personally undertook the preparation of a new incentive-type contract, working directly with General Luedecke, Hughes officials, and Surveyor contracts personnel. This was a very difficult task, partly because of the sensitivity involved in determining the status of contract activities at the time. Things were generally fouled up, and there were several loose ends that could probably be attributed to inertia: a number of technical changes had been made but never incorporated in the contract, and it was hard to tell who was responsible for what. By late 1964 about 46 modifications and 80 change orders had been accumulated. Not until these had been negotiated could the combination of NASA Headquarters, JPL, and Hughes' top-level management reach agreement on how to proceed. The revised contract was finally hand-delivered to Hughes by General Luedecke on the day the first launch occurred and was signed by Hughes officials just hours before liftoff. The signing signaled the end of a tumultuous period of planning, reprogramming, and recovering from a jumble of technical and management problems.

Another essential person in the Surveyor success story was Robert Garbarini. Bob had been serving as Chief Engineer for the Office of Space Science and Applications, and when Surveyor got into trouble, he pitched in to help in the program reviews and technical recovery planning. His almost full-time concentration on this freed me from many of the technical management matters I had been overseeing; his strengths and experience plus his wonderful attitude in dealing with people were major factors in the turnabout of Surveyor. Bob worked closely with Ben Milwitzky, who had long provided technical strength in the program management area. The combination of Garbarini's management capability and Milwitzky's thorough technical knowledge of the spacecraft, people, and status of all the hardware and tests provided a powerful combination for working with JPL and Hughes after NASA and JPL management finally got together.

Although continuing to direct the Ranger, Lunar Orbiter, and Mariner activities, I remained involved in the Suveyor program, issuing directives for action, making assignments, and working on special problems like the one concerning the vernier engine contract. It is painfully clear now that the Surveyor program was in deep trouble in early 1964. Fortunately, "all the king's men" rallied to the cause and were successful in putting it back together again.

One of the serious incidents I now recall with a smile was related to a rash of human errors occurring in the Hughes Aircraft Company during this time. Intense management attention was directed to the problems at regular monthly meetings involving NASA Headquarters, JPL, and Hughes officials. Bob Garbarini, Ben Milwitzky, and I from NASA Headquarters and General Luedecke, Bob Parks, Gene Giberson, and Howard Haglund of JPL were usually involved. Hughes officials included Pat Hyland, Allen Puckett, John Richardson, Fred Adler, Bob Sears, and a newly named Hughes project manager, Bob Roderick. At these meetings we reviewed all aspects of the problems and assessed progress and plans so that immediate attention could be given to recovering from our series of misfortunes.

Mindful of morale and the value of incentives to encourage thoroughness and good performance, we employed the "carrot and stick" management approach. The "carrot" took the form of incentive awards for meeting schedules, maintaining costs, reducing the number of man-hours involved, and so forth. As a "stick" to help reduce human error, I provided a means for Hughes management to recognize those who had caused errors or made visible mistakes. The idea came to me from my Army days when, during target practice, GIs in the pits beneath the targets used flags for signaling the results to the firing line. The most widely recognized of these was a red flag, known affectionately as "Maggie's drawers," which signified a miss of the entire target. I had a large "Maggie's drawers" flag made up, paid for it out of my own pocket, and sent it to John Richardson, a vice president of Hughes, with a letter suggesting that when a major incident occurred during the test or fabrication of a Surveyor the flag be flown from the company's flagpole for all to see. I also recommended that the group causing the incident be given a sign in their work area of the plant to help others recognize them as having "pulled the boo-boo" that merited the flag.

To my knowledge the flag never flew from the Hughes flagpole, but, based on the grumbling we heard and the fact that the quality of the test activities improved, I believe it served its purpose as a spur to avoiding human error. After the project was complete and everyone was relaxed again, John sent a nice tongue-in-cheek letter, thanking me for the help, and returning the flag so that I might use it on another "worthy" project. Of course, after the remarkable success of Surveyor, I would have been happy to accept it had the flag been returned with a punch in the mouth.

After so much has been said about the development and management aspects of Surveyor, it is time to recall the exciting events of the missions and the scientific findings. As initially planned, Surveyor spacecraft were to have elaborate payloads including seismometers, X-ray diffractometers and spectrometers, drills, and a soil processor that was to receive material from a soil mechanics surface sampler. Launch vehicle constraints reduced the payload to only 63.5 pounds on the first mission, and the instruments were pared down to a TV camera and some engineering measurements that made use of the landing gear structure, temperature sensors useful for other purposes, and the landing radar data interpreted for measuring reflectivity. By judiciously instrumenting the spacecraft, it was possible to deduce lunar surface mechanical properties, thermal properties, and electrical properties.

To disappointed scientists, this payload was unworthy; but compared with the pioneer who had only his eyes, Surveyor was well equipped. In addition to data obtained from engineering instrumentation, the camera produced and recorded image information about lunar topography, the nature of the surface, the general morphology and structure, the distribution of craters and debris on a fine scale, and, from observations of the footpads, an idea of the bearing strength. It also served as a photometer, giving for the first time a correct photometric function to compare with telescopic observations.

Surveyor 1 was launched from Cape Kennedy May 30, 1966, on a direct-ascent lunar trajectory. Approximately 16 hours after launch, a successful midcourse correction maneuver was executed, moving the landing point some 35 miles, to an area north of the crater Flamsteed in Oceanus Procellarum. Because telemetry indicated that one of the two omnidirectional antennas may not have fully deployed, a terminal maneuver was used that assured communications during descent. The spacecraft properly executed all commands, and the automatic closed-loop descent sequence occurred normally. Data indicated that the touchdown velocity was approximately 10 feet per second.

Of course I was in the Space Flight Operations Facility at JPL for the landing, along with other Headquarters associates and Congressman Joseph Karth. Based on experience, we had no right to expect success on the first mission, and I was prepared for the worst as telemetry reports came in. The main retro rocket had fired. The radar had locked onto the surface, and the verniers were thrusting. The spacecraft attitude was stable, and then came the altitude callouts: 1000 feet . . . 500 . . . 50 . . . 12 . . . Touchdown. I could hardly believe it, but then, before long, the first pixels of a TV frame showed the footpad on the surface.

Within a few hours we knew a lot about the Moon. The 596-pound craft had rebounded slightly after touchdown, its footpads pushing the surface material outward slightly. The evidence was clear that the Moon's surface was strong enough to support Apollo, and the topography that had accepted a Surveyor appeared hospitable for a manned spacecraft as well. Shaking hands with Congressman Karth as we celebrated the success brought flashing memories of Mariner 1 and the Ranger failure reviews we had shared—this moment was sweeter than sweet.

The early success of Surveyor 1 was the stimulus needed to charge ahead. By the time Surveyor 3 would be launched, a sampling scoop on an extensible arm could be added to dig in the soil, test the hardness of the material, and see how it behaved in a pile. This ability to manipulate the surface the way a person might with his hand would add another dimension to exploring.

The formal name for the sampler scoop was Soil Mechanics Surface Sampler (SMSS), and its conceptual originator was Professor Ronald Scott of the California Institute of Technology. Floyd Roberson, a JPL engineer who worked with Scott, was to be the operator of the arm and its scoop, and it was his honor to command it to dig the first trench on the Moon. This was done using the TV camera to see what was being achieved a step at a time. The camera could not look directly at the surface, but at a rotating mirror. Roberson had to learn to operate the arm with every movement reversed because of the mirror; this he did by working with a laboratory model of the arm in a sandbox.

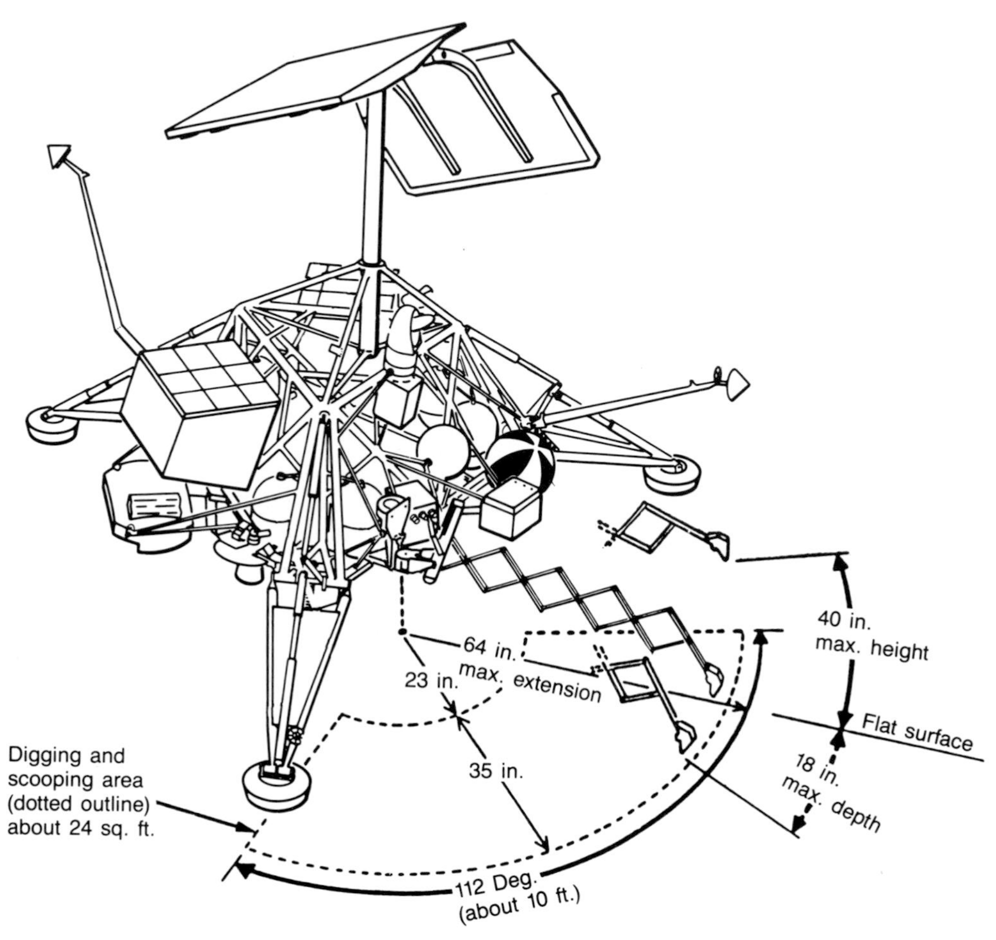

Surveyor surface sampler movements

Surveyor 2 failed during midcourse maneuver when one vernier engine did not ignite; Surveyor 3 landed in Oceanus Procellarum 390 miles from Surveyor 1 with the soil sampler aboard. While Surveyor 1 had placed man's eyes on the Moon, Surveyor 3 added an arm and a hand to work the surface. Its engines did not shut off before touchdown as planned; as a result it made two "touch and go's" before coming to rest at an angle of 14° below the rim of a small crater. Examining the footprints gave much insight about the nature of the surface.

By manipulating the sampler, Roberson conducted eight bearing strength tests, pressing its flat side against the surface. He did impact tests by dropping it, and dug four trenches in the cohesive soil. However, the most exciting use of the arm was to help examine a typical "object" lying nearby.

The object looked like a small white rock, but until now there was no way to be sure what the consistency of the object might be. After 90 minutes of manuevering inch by inch, pausing for television verification of each step, Roberson approached the object with the scoop jaw open. Careful not to miss, thereby pushing the object away, he closed the jaw, enveloping the sample and lifting it from the surface. Then came the test that could break an Earth-made brick: the jaws were commanded to exert pressure of 100 pounds per square inch on the sample. It did not break.

Hal Masursky, a scientist from the U.S. Geological Survey, was ecstatic. "If you can't crumple it like a soft clod, dig it with your fingers, or break it with a pressure or a whack, it must be a rock." In spite of this strong feeling, he cautiously described the sample as "highly consolidated material."

For Surveyor 5, the climax of the mission was the first chemical analysis of lunar material. It was done by an instrument 6 inches on a side, designed by Anthony Turkevich of the University of Chicago. Lowered to the surface by a cord, the alpha back-scatterer bombarded atoms in the soil with helium nuclei (alpha particles), knocking out protons and scattering back alpha particles and protons to a detector. By the number and energy of the particles scattered, Turkevich deduced the soil's composition. To the surprise of some scientists, the three most abundant elements were oxygen, silicon, and aluminum, in that order—the same order found in Earth's crustal materials.

Surveyor 6, carrying the same type of payload as Surveyor 5, touched down in another potential Apollo landing site, performed a confirming analysis of the soil, took 30 000 pictures, and performed the first rocket flight from the surface of the Moon when its vernier engines were reignited and allowed to "fly" the spacecraft some 8 feet to a new location. After this bold venture, we were ready for a real test. Besides, four potential Apollo sites had been found suitable, and it was time for the scientists to call the shots.

The site chosen for Surveyor 7 was in the rugged highlands among ravines, gullies, and boulders just 18 miles from the rim of the bright crater Tycho. After a look around at the "exciting" terrain, Turkevich's instrument was to be lowered to the surface, but it failed to drop. Roberson and his remote arm were brought into play, and gently lifted it to the surface. After one series of measurements, Roberson then dug a trench and moved the instrument to the freshly exposed soil at the bottom for another analysis. Finally, he lifted the instrument and placed it atop a rock. The sampler was also used to shade the instrument from the hot Sun. In addition to helping its scientific colleague, the sampler picked up and pushed clods, hit and weighed rocks, dug other trenches, and made more bearing tests. The way the men, the camera, the arm, and the dry chemistry instrument worked as a team illustrated the power of a partnership. This last mission put the frosting on a scientific expedition to Earth's nearest neighbor and was a turning point in automated spacecraft applications.

Surveyor was a once-in-a-lifetime experience. In addition to the wonderful opportunity to land sophisticated spacecraft on the Moon, the trials and tribulations during the effort promoted the maturing of such undertakings. It might be presumptuous to say that Apollo engineers and officials were able to proceed with greater confidence because of a prior successful automated venture, but this may have been the case. I know for certain that the Surveyor experience bonded a group of us Earthlings together in a way that nothing but struggling and succeeding as a team can do.

The Back Side of the Moon

Lunar Orbiter was a less than imaginative name for one of NASA's most successful automated projects. After all the hoopla that followed the naming of Ranger, I was somewhat dismayed that a generic label like Lunar Orbiter was affixed to the lunar photographic mission. This was the name used by Langley engineers who were defining the project, and although I registered concern with Ed Cortright, who had been responsible for studies and agency policies recommending logical evolutionary names, he decided that it would be better to go along with the new project team. Names like Pioneer, Mariner, Voyager, Surveyor, and even Ranger sounded to me like space exploring machines; Lunar Orbiter had all the romance of calling a favorite pet "Pet."

But even with its unexciting title, the Lunar Orbiter project became a sweeping success, accomplishing all its primary goals and then some, with only minor hitches to stir up excitement for the project team. All five orbiters completed useful lunar photographic missions. There were no launch vehicle failures and no major spacecraft failures. The first three missions satisfied the primary purpose of the program, which was to photograph proposed landing sites for manned Apollo missions. The final two flights were devoted largely to broader scientific objectives; photographing the entire near side of the Moon and completing coverage of the far side. The five orbiters together photographed 99 percent of the Moon, including the side away from Earth, which had only been vaguely visualized by the Russian Luna 3.

As mentioned earlier, the Surveyor program was originally defined and initiated to include both an orbiter and a lander. A common set of basic hardware was to provide Surveyor landing spacecraft, which would obtain lunar data from the surface of the Moon, and Surveyor orbiters, which would map the Moon and provide overall coverage from orbit. The orbiters were to use the same basic airframe components as the landers but with the landing gear removed and different retro motors designed for placing the spacecraft in orbit. Of course the instrument packages of the two complementary spacecraft would have been different to address the different objectives of orbiting and landing missions. The two types of Surveyor were to serve as a team and produce orbital reconnaissance information plus "local site" landing data, thereby greatly increasing our total knowledge of the Moon through the synergistic benefits of broad and close-up coverage.

Although the early Surveyor spacecraft specification recognized the orbiter from the outset, initial design emphasis was given to the landing craft; it was reasoned that the orbiting requirements could almost be considered an extension of the cruise mode. JPL's involvement in getting the lander vehicle defined and designed, plus their burden with the Ranger and Mariner projects, made it difficult for them to assign people to work on a Surveyor orbiter. When the definition of the orbiter did not materialize, though both the Surveyor lander and the Surveyor orbiter had been approved by NASA and authorized by Congress, it was evident that something else had to be done if we were to get the combination of orbital and surface information that was needed to support the Apollo mission.

During a senior council meeting of the Office of Space Sciences and Applications in January 1963, I asked Floyd L. Thompson, Director of the

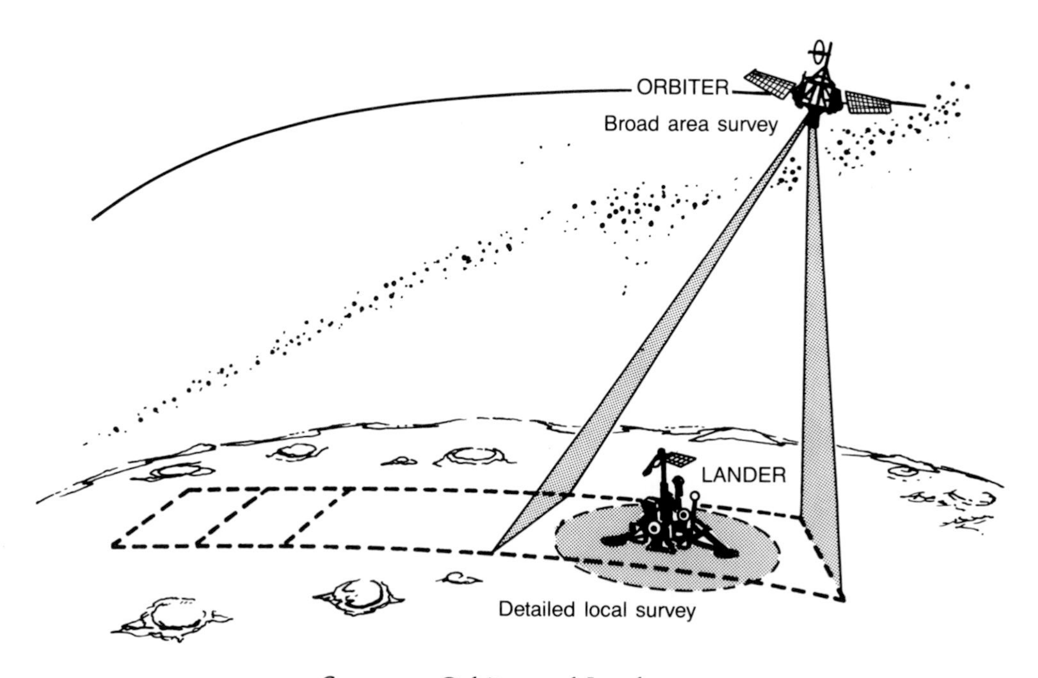

Surveyor Orbiter and Lander team

Langley Research Center, if Langley would be willing to study undertaking a lunar orbiter effort. Earlier Langley had been interested in lunar experiments on Ranger missions, and I knew this research center to have personnel with talent and capability who were not presently engaged in space projects. Thompson agreed to explore such a possibility; thus began an activity that led to the Lunar Orbiter project.

It was clear that many things needed in the study of the Moon were offered by orbiting reconnaissance. First, planning for Apollo missions had by this time narrowed possible landing sites to a zone on the near face of the Moon bordered by $\pm 20°$ latitude and $\pm 45°$ longitude. Detailed surveys of the area would obviously be important to final selection of landing sites, and high-resolution photography would, of course, provide the maps needed for navigation and final touchdown.

A spacecraft orbiting the Moon would also have obvious scientific potential for photographing the back side, something that had never been done with Earth-based telescopes, because the same hemisphere of the Moon always faces Earth. A great deal of conjecture existed about the back side and the nature of its surface, in spite of the fact that the Russians had obtained some low-resolution photos. In addition to extremely significant scientific information, an orbiting spacecraft could also provide a new perspective on the Moon as a planetary body. This, in conjunction with Ranger and Surveyor data, would add greatly to our knowledge of Earth's neighbor and would allow the formation of specific questions of major interest for manned lunar missions.

After Langley's participation was approved, steps were taken to develop a project with this fresh team. By the time the Lunar Orbiter project was formed, some 2 years of instructive experience with Rangers, Mariners, and Surveyors at JPL had taught us a lot about the management practices needed for major projects. After initial difficulties, NASA Headquarters had evolved a system of working with field centers that was spelled out in a management instruction first published in 1961 and revised in March 1963. As I was part of the team developing this policy and implementing it at JPL, the policies and procedures embodied in NASA Management Instruction (NMI) 4-1-1 were fresh in my mind during the period when we were negotiating with Langley on the Lunar Orbiter project.

The document was relatively straightforward in establishing the hierarchy of responsibilities for NASA Headquarters and for field centers engaged in project activities. In general, it summarized Headquarters' four basic responsibilities: (1) establishing objectives, (2) scheduling the milestones (considering technical, fiscal, manpower, and other requirements), (3) budgeting and obtaining the required financial resources, and (4) seeing that projects were properly implemented and carried out in the field.

Field centers were assigned project management responsibilities, with project managers having principal authority for implementing the work. In addition, definitions were given for assignments of system managers who would report to the project manager and be responsible for each major system such as the spacecraft, the launch vehicle, the tracking and data acquisition system, and spaceflight operations. Although the concept of making vertical and horizontal assignments among centers was initially quite controversial, it was not long before the organization of projects and definitions provided by this management instruction were understood and accepted. It was significant that such a clear framework existed for negotiations between our Headquarters office and Langley at the beginning of this project; it was easy to reach agreement on organizational matters and to get on with the job. It had not been possible to do this readily during the evolution of project management activities at JPL, when many different patterns of operation were proposed.

The man assigned direct responsibility for managing the new Lunar Orbiter program at Headquarters was Lee R. Scherer, a very capable naval officer who had been assigned to NASA for a 1-year tour of duty. He was an honor graduate of the Naval Academy and had served as an AED, Navy code for aeronautical engineering duty. His assignment was intended to provide experience in space activities that would help him be of value to the Navy. About the time his tour with NASA was to end, the Navy role in space was curtailed by the Secretary of Defense. When faced with a probability of continuing his Navy career without much hope for involvement in space, he opted to retire and join the Lunar and Planetary Program Office. This was a timely decision for NASA, as he did an excellent job and smoothly guided the many activities of the Lunar Orbiter.

Lee was a very outgoing person, at home with officials, scientists, engineers, and laypersons. His skill in coordinating interface matters between Headquarters, Langley, Lewis, JPL, and the many contractors was a significant factor in facilitating technical progress, even though he did not seem to become too technically involved.

He came to work one day in a bright blue and gray plaid sport jacket that I thought was good looking, but admittedly it was not quite in keeping with the dress usually worn by Headquarters officials. He received a lot of ribbing because of this "race track" jacket, but when he wore it at the Cape during the first launch operation and then at JPL during the entire mission operation, it became the project's good luck symbol for success.

Although the already proven Atlas/Agena vehicle was to be used for the Lunar Orbiter, a new and somewhat worrisome aspect of launch vehicle integration became apparent. The role of the Lewis Research Center in procuring, integrating, and launching vehicles had just been expanded to include overall responsibility for the Atlas/Agena, and a new team had been assigned to manage this system. By NASA ground rules, the launch vehicle system manager reported functionally to the Langley project manager, but an age-old rivalry between these former NACA research centers made this relationship somewhat sensitive, especially since both project groups were newly assigned and anxious to prove their mettle. Questions arose about interface matters, such as who should design and procure the interconnect hardware between the Agena and the spacecraft, or who should be responsible for the shroud that protected the spacecraft but also attached to and separated from the vehicle. Lee and I found ourselves in the role of moderator several times. However, in spite of a few delicate and potentially volatile situations involving the two organizations, all vehicle interface and development matters worked out well. For once, all the launches were successful.

Beginning from scratch with a new project, Floyd Thompson chose Clifford Nelson as Project Manager and assigned a few outstanding engineers to work with him in the development of plans. Cliff had recently managed a smaller project called Project Fire involving a rocket-launched reentry probe at Wallops. Because at the time Lunar Orbiter was the only major space project at Langley, Thompson and his deputy, Charles J. Donlan, maintained close cognizance over activities and imparted a considerable amount of experience and wisdom to the process. Assignments were made to old-timers like Israel Taback, Ed Brummer, and Bill Boyer and to newer faces like Cal Broome and Tom Young. All five did great jobs on Lunar Orbiter and were destined to become giants in the Viking project. From my Headquarters point of view, dealing with this new team that had a very cooperative outlook was a real pleasure—quite a different experience from the struggles during the start-up of the Ranger project.

In addition to the new NASA team, a group from the Boeing Company that was new to NASA became the contractor to develop Lunar Orbiter spacecraft. The Boeing group had become "available" to prepare the orbiter proposal when a large Air Force project called Dynasoar was canceled. This left an almost intact nucleus of first-rate engineers to work on Lunar Orbiter, providing the talent needed to establish a complete team almost immediately. Boeing won the competition with a concept somewhat different from that employed by most proposers, offering a three-axis stabilized spacecraft with more capability than I had initially envisioned as necessary for the task. All in all, the combination of factors existing at Langley and at Boeing during the establishment of the project was unique and undoubtedly contributed to its success.

The timing of project initiation was also significant. The need for Apollo planning information was considered somewhat critical, and the recollection of difficulties in providing scientific payloads for Ranger, Mariner, and Surveyor led to a decision by the Office of Space Sciences and Applications that the orbiter mission would focus on a single purpose, namely, photography of the Moon. Because this objective involved detailed design tradeoffs between the camera system and the spacecraft, and because the scientific returns were to be used largely for Apollo mission support, camera systems design and photographic mission planning were defined to be "engineering" activities, with "support" to be provided by the scientific community, rather than the other way around. This focused decision-making responsibilities for the scientific payload equipment within the project office, facilitating payload-spacecraft integration to a greater degree than had been experienced with other missions.

Another ground rule adopted by the project office was that proven hardware from any source would be integrated into the orbiter if possible. Langley and Boeing engineers immediately reviewed all information on existing systems that might be applicable to a Lunar Orbiter mission, including techniques for attitude stabilization and control, midcourse correction, and maintaining the housekeeping functions of power, communications, temperature control, and the like. Even the camera system configuration that was chosen had been used on Earth-orbital flights. It was in fact a derivative of a camera developed for military reconnaissance that had been superseded by equipment with greater capability, but also having such a high military classification that Department of Defense personnel did not wish to see it used in NASA's open society. This use of proven technologies and equipment allowed Langley and Boeing to place emphasis on new developments required specifically for this mission.

The matter of photographing the Moon was challenging, partly because of the unusual photometric properties of the lunar surface. From Earth-based observations it was known that the reflective properties of the Moon are quite different from those of Earth; this and the fact that the Moon has no atmosphere for light scattering means that objects within shadows are invisible. From a study of these lunar characteristics, it was determined that much of the photography should be obtained during morning Sun at angles of 15° to 40° above the local horizon. This would produce a reasonable balance of shadows so that topographical features would stand out.

Since the region of the Moon near the equator was of prime interest to Apollo planning, it was targeted for initial missions to ensure that all the necessary photos would be successfully obtained with only five spacecraft. The first requirement for proper lighting conditions—the angle of the Sun with respect to the region being photographed—was satisfied by launching at a time when arrival at the Moon would find the Sun's morning rays making the proper angle. The Lunar Orbiter's trip time to the Moon was 90 hours, because the flight was planned as a near "minimum-energy" trajectory to reduce the amount of retropropulsion required for establishing orbit. Of course, the lighting changed as the phase of the Moon changed, so any given mission had to be scheduled to allow the photographic sequence to progress along the surface ahead of the day-night terminator as the shadow moved. Once lunar orbit was established, its geometry would permit the orbiter to maintain an essentially fixed orientation in inertial space relative to the Moon, so that waiting in orbit would allow the rotation of the Moon on its axis to bring the targets of interest under the low point of the spacecraft orbit. Refinements in the orbit were possible by additional burns of the retro rocket, but because of the risk involved they were kept to a minimum.

When the targets were favorably located under the orbit, the spacecraft was reoriented from its solar power attitude to look downward and take a series of photographs. If coverage greater than that achievable on a single pass was required, blocks of coverage were built up by overlapping photography on successive orbits. The overlaps were defined in advance, depending on the cameras used, and, in addition, stereoscopic coverage was provided by the wide-angle, 80-millimeter lens system.

After most of the mapping photographs were taken in direct support of the Apollo requirements, a number of available periods resulted in photographs of great scientific and general interest, including oblique views and pictures of regions other than those thought to be of immediate interest to Apollo. The first three Lunar Orbiter missions were successful in obtaining all the necessary Apollo site coverage, with some verification of earlier data and a complete series of maps for the region of interest. On missions 4 and 5, higher orbital inclinations were ordered, and maximum scientific coverage was provided from these two spacecraft. Because of the launch successes and operational successes of the spacecraft, Lunar Orbiters returned scientific data that we had not dared to anticipate at the outset of the program.

The spacecraft was a three-axis stabilized vehicle, weighing about 850 pounds at launch by the Atlas/Agena. Electrical power was provided by four solar panels, with batteries for a limited electrical load during periods of sunset or when the spacecraft was oriented for photography. Two antennas, one a high-gain directional and one omnidirectional, provided communications with the spacecraft in the same general manner as for Rangers and Mariners. Thermal control for the vehicle was primarily passive, with a limited number of electrical heaters. The attitude references for yaw and pitch were provided by Sun sensors so that the solar panels faced at right angles to the Sun. The roll axis reference was provided by an electro-optical sensor that tracked the star Canopus. The high-gain antenna pointed toward Earth with the assistance of a rotatable boom on the unit which could be programmed. The spacecraft normally maintained this Sun-Canopus oriented attitude control, except when it was reoriented to align the rocket engine for midcourse correction, for lunar orbit injection, or during periods when the cameras were being pointed and the spacecraft was reoriented to allow photography.

Most functions of the spacecraft were controlled by an onboard programmer. This unit received commands from Earth stations and either executed them immediately or stored them for execution at a later time. Sufficient memory was available in the system to allow automatic control of the spacecraft functions for a period of several hours.

In addition to the photographs of the Moon, two other forms of information about the space environment were provided. A group of micrometeoroid detectors was located in a ring just below the fuel tanks to record punctures by micrometeorite particles. Each detector was a pressurized can which, when punctured, would send a signal to Earth so that both the event and the location of the impact could be determined. Two proton radiation detectors were carried to allow evaluation of the environment affecting the film. Shielding on these detectors approximated that at two critical locations within the photographic system, and telemetered dose rates allowed evaluation of the fogging effects any solar proton event might cause. These proton events were of serious concern, and the latest information on solar flare activity was always factored into prelaunch planning.

In a respectful sense, we spoke of the photographic system as a pair of sophisticated "Brownie" cameras housed in a pressurized container, along with a developing system somewhat like that used in laboratory processing of film, except that liquids were contained in webbing material instead of pans. The entire photographic system was housed in a thin aluminum shell maintained under pressure between 1 and 2 psi, with a high-pressure supply of nitrogen available to maintain this pressure in the event of small leaks. Temperature control within the unit involved a mounting plate with fins to radiate heat from the underside of the shell, plus automatically controlled heaters of the electrical resistance type. Temperatures were controlled within $\pm 1^\circ$, and humidity was maintained at 50 ± 10 percent with the help of potassium thiocyanate pads.

The camera lens had to be protected from the cold of space by an insulating door, in appearance like those constructed by the trap-door spider. This light and somewhat flimsy structure was recognized as a success-critical item when it failed to open during thermal vacuum tests before the first flight. This was the only failure in a series of systems tests, but it was enough to delay shipment and necessitate a rework and retest before the first flight.

In spite of special attention given to the thermal door problem, a failure did occur during the fourth mission; the door did not close after a photographic sequence. Fortunately, it had been designed to allow use in a partially opened state for temperature control, and it was possible to give it step commands that would move it a notch at a time. This command mode was used to save the mission, but there were a lot of worried people and a myriad of commands involved in the process. The incident was a frightening reminder of the small links in the chain that were critical to success, many of them easily overlooked during the development of a complex set of high-technology items, but each as important as the most sophisticated element.

The photographic system was composed of three basic sections: camera, processor, and readout equipment. Of course, many interconnections were necessary to make the system operate as a unit and react to commands transmitted from the ground.

The film was Eastman Kodak type SO-243 High Definition Aerial Film, 70 millimeters in width. As the film was pulled from the supply, it passed first through the focal plane of the 80-millimeter lens, sometimes referred to as the wide-angle lens. This lens-shutter assembly was an off-the-shelf unit modified from f/2.8 to f/5.6 with a Waterhouse stop. Modification also included elimination of shutter speed settings, except 1/25, 1/50, and 1/100 of a second. A neutral density filter was added to the lens to help achieve a balance in exposure with the 610-millimeter lens. Simultaneously with exposure of the 80-millimeter format, exposure occurred on the 610-millimeter lens systems, and a 20-bit code showing the time the photograph was taken was exposed adjacent to the 80-millimeter format. The 610-millimeter lens, modification of an earlier design by Pacific Optical, used a folding mirror and a focal plane shutter to expose a format of approximately 5° by 20° versus the 80-millimeter format of approximately 35° square. Following each exposure, the film was advanced exactly 29.693 centimeters (11.690 inches). This brought the last 80-millimeter frame to a position just short of the 610-millimeter platen, bringing fresh film onto both platens and readying the system for the next exposure. In this manner, the 80-millimeter and 610-millimeter frames were interlaced on the same strip of film. A pre-exposed pattern of Reseau crosses was present on the film for indexing, along with a nine-step gray scale, power resolving targets, and reference numbers.

Coming closer than about 200 kilometers to the lunar surface made image motion a significant degrading factor because of the speed of the spacecraft over the surface. Therefore, image motion compensation was provided for both lens systems, since some of the high-resolution photographs were to be taken at an altitude of only 20 kilometers. To accomplish this, a portion of the field of view from the 610-millimeter lens was fed to the velocity/height sensor located physically above the camera plane. This optical signal was analyzed by the sensor, time correlated, interpreted, and transmitted into a servomechanism output used to drive both camera platens so as to null the image motion. In other words, the film speed was adjusted by the image motion compensation sensor system to compensate for the motion of the spacecraft past the target area. Since the camera could take up to 20 photographs in rapid succession at framing rates as high as 1.6 seconds per photograph, buffer storage was provided for the film between the camera and processor. Film was pulled through the camera by the film advance motor and temporarily stored on a camera storage looper system which could hold up to 21 frames of exposed film before sending it through the developer.

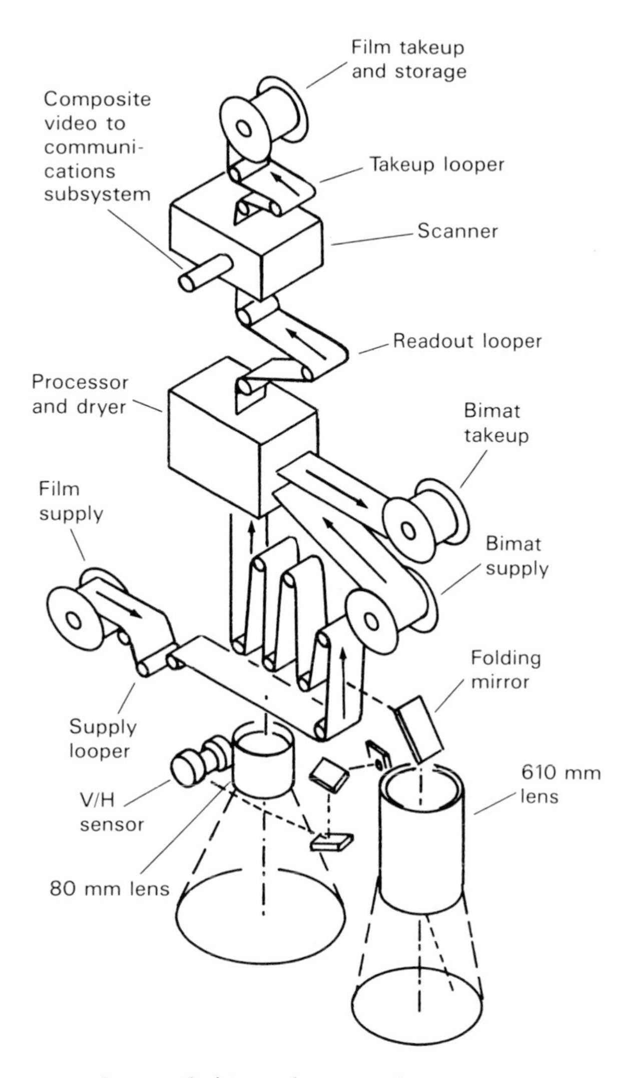

Lunar Orbiter photographic system

After completion of a photographic pass, the processor was turned on. Film went into the processor and was laminated with Eastman Kodak type SO-111 Bimat film presoaked with Imbibant type PK-411. Processing of the film to a negative took place during travel around the processing drum at a controlled temperature of 85° F. After processing, the film and Bimat were separated. The Bimat was discarded into the Bimat takeup chamber, and the film passed over the dryer drum, where it was subjected to a temperature of

95° F, and moisture was driven off for absorption by pads around the periphery of the drum. Following drying, the negative film was twisted once again through 90°, passing out of the processor and into the readout looper, which was similar in principle to the looper after the cameras, before processing. At this point, the readout looper served only a control function, being partially filled with processed film, then signaling the motor on the takeup spool to empty it once again. In this manner, the photographs were exposed, processed, dried, and stored, ready to be transmitted to Earth.

After all the film had been processed, the Bimat developer was cut free by a hotwire device, making the processor free wheeling so that the readout could proceed until all data had been examined. The selected readout mode made data available at any time and was limited only by the capacity of the readout looper. Two readout modes were possible. In normal operation, only selected readout was conducted prior to completion of all the photography and processing. After processing, the readout could begin from one end and continue until all the film had been read. Film travel during readout was opposite to the direction of picture taking; thus the last pictures obtained would normally be accessible first for readout.

The readout concept involved a light scan generated by a linescan tube. Images were fed through optics to a photomultiplier tube which in turn fed a video amplifier and transferred the signal into a 0 to 5 volt, 0 to 240 hertz video signal for transmission to Earth. In the readout assembly, the film advanced in 2.5-millimeter segments. During a 23-second pause between advances, the film was clamped in the readout gate and scanned with a raster of about 287 lines per millimeter. Light for this scan was generated by the linescan tube, which provided an 800-hertz horizontal sweep of an electron beam across a revolving phosphor drum anode. The resulting flying spot, approximately 200 microns in diameter, was "minified" 22 times and imaged on the emulsion side of the film. The vertical component of the raster was generated by moving the minifying lens or scanner lens at a precise rate across the film. After scan of each segment, the film was advanced, the lens reversed, and the next segment scanned in the opposite direction.

Light transmitted through the film was collected by optics and fed to the photomultiplier tube for conversion to electrical signals. The film was thus read out in "framelets," each 2.5 millimeters by 65 millimeters and each requiring about 23 seconds to transmit. One frame, defined as one 80-millimeter and one 610-millimeter exposure pair with their associated time-code data, required 43 minutes for transmission.

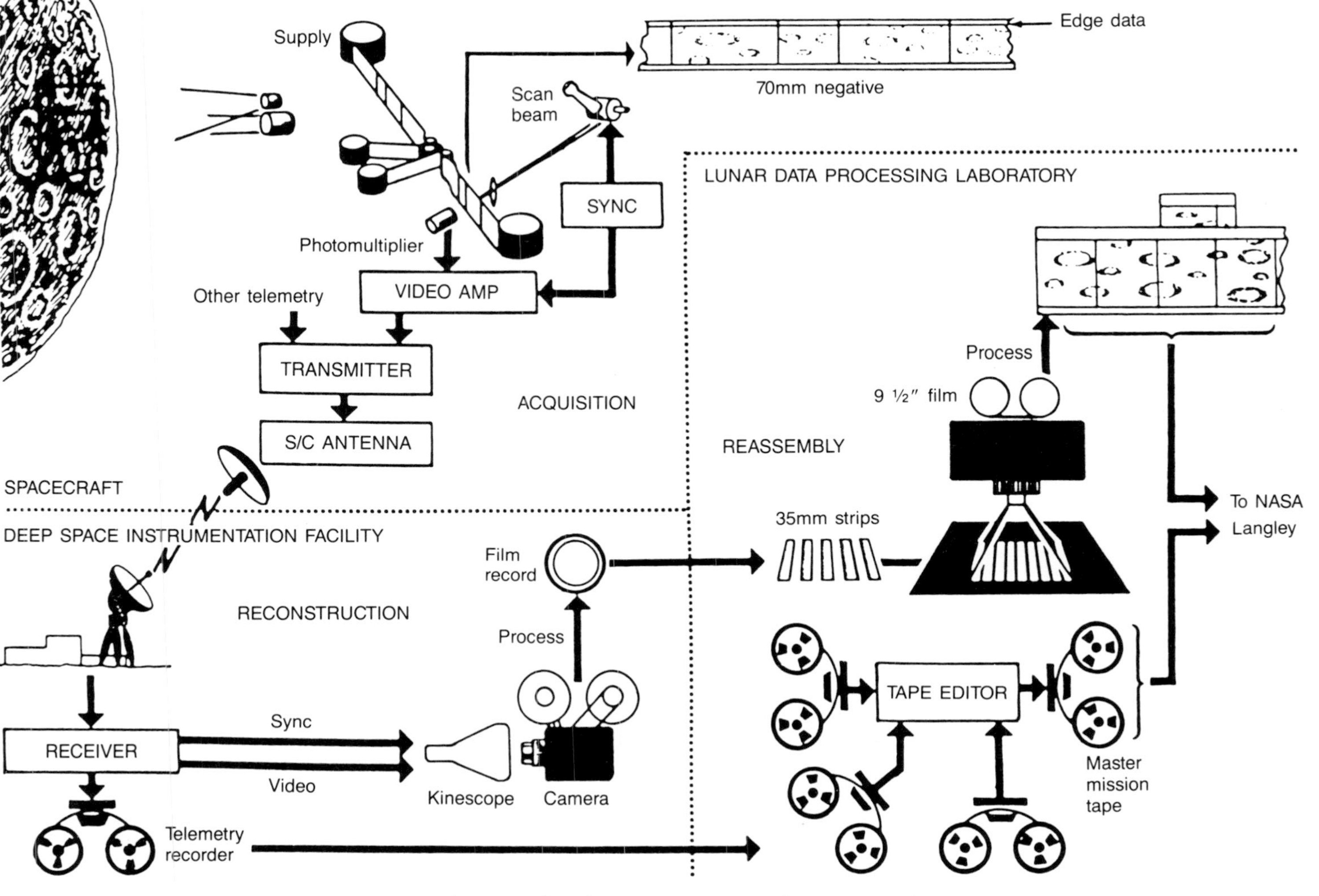

Lunar Orbiter photographic data acquisition and reconstruction systems

After receipt and processing at one of the Earth stations of the Deep Space Network, the video signal was fed to one of two recording devices. Predetection video recordings were made of each readout sequence handling on Earth. At the same time, the ground reconstruction electronics were used to regenerate the signal. This essentially provided the reverse of the spacecraft readout, taking the video signal and driving a kinescope whose linescan was imaged onto moving 35-millimeter film; thus, each framelet 2.5 millimeters by 65 millimeters in the spacecraft became a framelet 20 millimeters by 420 millimeters on the ground. These framelets were then laid side by side to provide reconstruction of all or part of the frames as they existed in the photographic system in lunar orbit.

Operations for the first Lunar Orbiter mission were conducted with deadly seriousness, and, as might be expected during the first flight, a number of anomalies occurred. The image motion compensator did not work properly, so that no extremely high-resolution photographs of any value were obtained. In spite of these troubles with the photographic system and problems with the orbiter attitude control system, a total of 205 exposed frames resulted. Of these, 38 had been taken in the initial orbit and 167 after the orbit had changed to provide the closer approach. The spacecraft did photograph all 9 potential landing sites for Apollo and, in addition, took pictures of 11 sites on the far side of the Moon plus 2 Earth-Moon pictures.

The pictures of Earth from the vicinity of the Moon that showed the lunar surface in the foreground were most spectacular and actually provided some new knowledge of the orbiter camera system capabilities, in addition to offsetting the losses in high-resolution data. The first of these Earth-Moon pictures was taken during orbit 16, about 5 days after the first photographs of the Moon were taken. Such pictures were not included in the original mission plan. They required a change in the spacecraft's attitude in relation to the lunar surface so that camera lenses were pointing away from the Moon. Maneuvering involved a calculated risk; the prospect of taking unplanned photographs of Earth early in the flight caused some concern among Boeing project leaders. Part of the concern was due to the fact that planning for the attitude control maneuvers and their execution had been completed during the high-activity period of the mission without much review and checking. If a problem had occurred because of this special picture taking that made it impossible to complete the mission as planned, the team would have been justly criticized.

However, the possibility of obtaining such interesting pictures led to a series of hurriedly held meetings among NASA program officials. Lee Scherer, Floyd Thompson, Cliff Nelson, Jim Martin, and I convinced ourselves that the photographs were worthwhile and then discussed the matter with the Boeing officials who were performing the mission under an incentive contract. It was not reasonable to modify the predetermined incentives; however, the contract allowed NASA officials to consider extra performance at the end of the program, and if the picture taking was successful, to justifiably reward the contractor for his extra efforts. As it turned out, Boeing officials agreed in spite of their project managers' concern, and the pictures were taken on two different orbits, 16 and 26.

Not only were these pictures spectacular from the standpoint that they provided the first view of Earth from the distance of the Moon with the Moon in the foreground, but they also gave valuable insight into the benefits of perspective shots of the lunar surface. Until these were taken, all pictures had been taken along axes perpendicular to the Moon's surface with the idea of providing map-like information. On subsequent Lunar Orbiter missions, however, oblique photography was planned and used. In talking with Neil Armstrong after his Apollo 11 mission to the Moon, I learned that some of the oblique photographs, which gave views of approach conditions like those the astronauts saw from their windows, were extremely helpful. It is interesting that these benefits may have accrued incidentally as a result of the early mission gamble.

Having been conceived with the primary objective of providing information essential for the Apollo program, it is fitting to note that the successful Lunar Orbiter also set the pace for achieving extraordinary performance. By the end of the third flight, objectives prescribed to support Apollo had been fulfilled. At the end of the fifth, the entire near side and some 99 percent of the far side of the Moon had been photographed. The resolution of the photography exceeded that available from telescopes many times; of course, the back side of the Moon has never been seen through Earth-based telescopes.

In addition to the excellent photographic coverage, new data were obtained about the size, shape, and mass distribution of the Moon. The major irregularities of the Moon's gravitational field were very significant discoveries, of interest scientifically and important to trajectory determinations of orbiting spacecraft. Micrometeoroid data and radiation levels in the vicinity of the Moon were also determined for the first time. While no surprises of significance were revealed from these measurements, "no news was good news" for Apollo planners.

While we never thought about conducting projects like Lunar Orbiter to learn how to manage, there were a number of good people who received excellent training on this project. Many of the principal Langley team members became key players in the successful Viking project a few years later. Jim Martin, who was hired to work on Lunar Orbiter because of his proven experience with industry, became a respected team leader and later guided the Viking effort as project manager. The team that conducted the Lunar Orbiter mission so well continued to distinguish itself as greater challenges were faced.

By the time the project was over, Lee Scherer's jacket had quite a few "mission hours" on it, for it had been in evidence at every major operational activity for all five flights of the Lunar Orbiter. Its symbolic contribution to the success of the program finally came to an end during a victory party at the Huntington Hotel in Pasadena, when the jacket was torn to shreds and divided among project members.

Tracking, Communications, and Data Acquisition—A Revolution

At an Apollo 11 victory banquet, master of ceremonies Joe Garigiola recited a "Yogi-ism" he attributed to his friend Yogi Berra: "If you don't know where you're going, you'll end up somewhere else." The truth in that bit of humor certainly applies to missions using unmanned spacecraft. Tracking and position determination are absolutely vital to the process of exploring distant targets, for it is essential to know where a spacecraft is and where it is heading in order to direct it to its destination.

Before the days of advanced radio communications technology, we would have been forced to use onboard celestial navigation principles—star trackers, sextants, and traditional navigation techniques—in conjunction with accelerometers to compute position, speed, and direction. The weight, power, and accuracy of such systems would have depended on a number of tradeoffs, and the complexities of developing and testing long-lived systems were many. Fortunately, advances in radio tracking technologies during and after World War II enabled us to use relatively simple, lightweight spacecraft systems in conjunction with large ground installations to obviate the onboard complexities of self-contained navigation systems.

A second essential in the unmanned spacecraft equation involves communications, both to deliver commands to a spacecraft in flight and to receive information about its findings in space. In the 1960s it was not possible to plan a journey of several months to a distant planet using only preprogrammed intelligence for the spacecraft; commands had to be planned from the outset. Since these spacecraft always went on one-way trips, they would have been of little use as emissaries for man if they had not been able to communicate their findings with accurate and interpretable information.

Thus, the pacing technologies for lunar and planetary missions included tracking, communications, and data acquisition capabilities. Position determination and communications functions were always combined, because the common elements of their radio disciplines bound their designers together.

For early space missions, specialized tracking and data acquisition systems were developed in parallel with each spacecraft and its instrumentation; however, it was soon recognized that it would be best if ground facilities for performing these functions could be designed and built to serve a number of projects. The concept of a Deep Space Instrumentation Facility (DSIF), combined with a Deep Space Network (DSN), emerged as a standard for meeting the needs of the many lunar and planetary missions and evolved over the years to support a spectrum of flight projects.

In summary, deep space missions require several basic types of support from the tracking, communications, and data acquisition facilities. For commands necessary to ensure control of spacecraft, there is a normal or routine capability, an emergency capability that usually is engineered to work under off-design conditions such as low power or unusual attitudes, and an emergency weak signal mode that allows searches for lost signals or for recovery from out-of-sync conditions. Radio navigation instrumentation is essential for determination of trajectories or orbits. This usually involves Doppler transponders and accurate pointing capabilities. For data acquisition, there are usually high and low bit rate modes to accommodate the differing requirements of continuously monitoring engineering data or interplanetary phenomena that do not vary rapidly, as well as the high bit rate requirements for imaging systems and encounter instrumentation. There are also special requirements for so-called radio science experiments that use radio signals and analyze changes in them caused by atmospheres and the interplanetary medium.

The amazing quality and performance of the NASA tracking and data acquisition systems cannot be recalled without giving credit to Edmond C. Buckley and Gerald M. Truszynski, who came to NASA Headquarters from Langley to lead the development of this enormous system. Ed Buckley was for a time Assistant Director for Spaceflight Operations under Abe Silverstein and was later the head of the Office of Tracking and Data Acquisition until his retirement in the 1970s. He came to Washington as a very experienced NACA engineer with an extensive background in telemetry and tracking system development and operations. One of his major efforts involved development of the Wallops Island range, a rocket launching, free-flight test facility that was built for free-flight transonic aerodynamics tests after World War II. Gerald Truszynski, who later replaced Buckley, had similar experience and continued the advance of capabilities, including tracking and data relay satellites and other innovations.

Although completely different by nature, Ed Buckley at Headquarters and Eb Rechtin at JPL respected each other and got along well. They provided an excellent example of headquarters and field center counterparts leading developments for a required technological base while managing programs and dealing with administrative chores. The telecommunications functions they provided were prime ingredients in the successful exploration of space. Just as impressive was their remarkable job of satisfying users' needs. From a program office point of view, working with these people—who never got their share of credit because of the supportive nature of their task—was indeed a pleasure.

Earth-orbiting satellites had been successfully tracked and interrogated by stations located within the United States, although one or two stations in the southern hemisphere helped. The fact that a low-altitude satellite came into view every 90 minutes made it relatively easy to track its location. Of course a satellite in Earth orbit was almost like a train on a railroad track; it tended to retrace the same general path in inertial space, orbit after orbit. For tracking lunar and planetary spacecraft, however, the process would be more like tracking celestial bodies, because a single station on the rotating Earth could see a distant spacecraft only during one-third of a day. This would not allow sufficient coverage to monitor critical functions and to transmit commands. Had ground stations been located only in the United States, very complex tradeoffs would have been necessary for timing events when the stations were in view of the spacecraft.

It did not take long for Rechtin, his principal system designer, Walter Victor, and the engineers at JPL to develop a plan for a network of three stations located approximately 120° longitude apart, so that one of the three would always be in view of any spacecraft. Obviously it was desirable for the principal station to be near JPL, if possible, and in the spring of 1958, a remote site suitable for a sensitive receiver (free from manmade radio interference) was located in a bowl-shaped area at Camp Irwin, an Army post some 50 miles north of Barstow, California, in the Mojave Desert. As this was a government reservation, it was not difficult to obtain approval to use this site. The problems associated with selecting sites and implementing plans for the other two network stations were more difficult, as one of the sites selected was in a dry lake bed near the Woomera test range in East Central Australia and the other in a shallow valley near Johannesburg, South Africa. Approval for these sites, of couse, required the Department of State to work out arrangements with the respective governments, including construction and staffing with nationals. During the 1960s, concern occasionally arose over the permanence of the South African site because of political unrest and the relationship of the South African and United States governments. In 1965, an additional site was established at Madrid, Spain, to backstop this uncertain condition and to ensure coverage near the Greenwich longitude.

The signals received by small radios with built-in antennas often come from commercial stations with as much as 50 000 watts of broadcast power. In the case of the early Ranger and Mariner spacecraft, only 3 to 4 watts of transmitter power were available. This placed a significant burden on the ground receivers to make sense out of the very weak signals. To acquire and sort out weak signals from random galactic background noise and manmade radio signals bouncing around Earth, tracking antennas had to be very large and highly directional. This meant that they had to be accurately steerable, for gathering the weak signal depended on their ability to focus on that single source. Most of the Earth satellite tracking antennas were driven by what were called "Az-El," azimuth-elevation drive systems, so that the coordinates were simply derived as normal and parallel to the surface of Earth at the point, and antennas were driven in two axes.

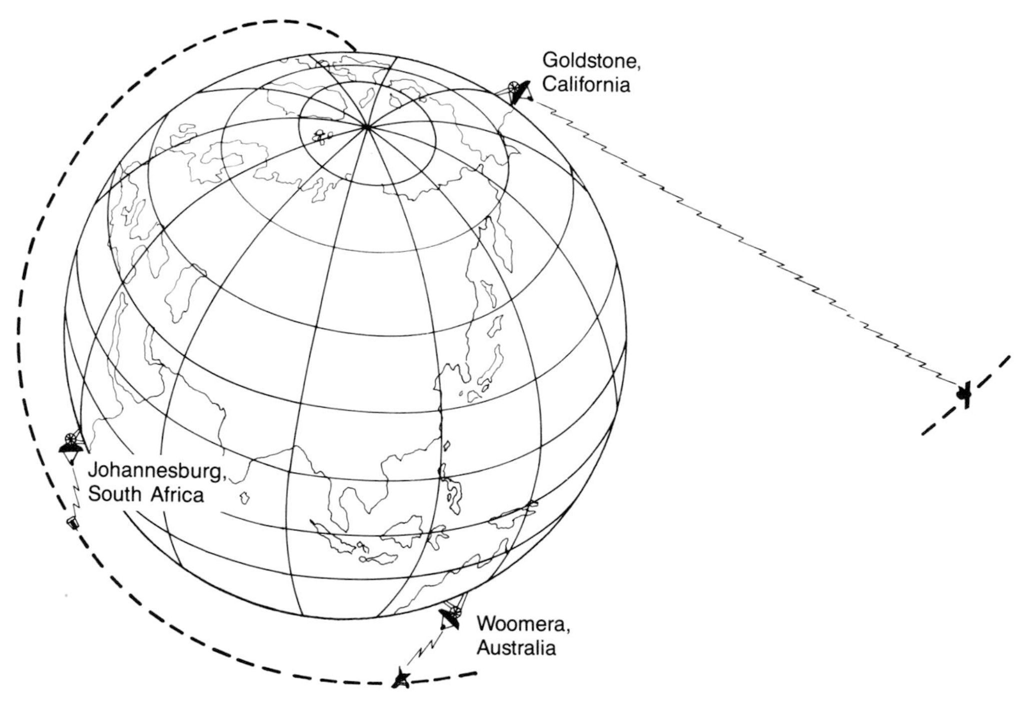

Location of Deep Space Network stations

Two principal members of the JPL staff, Robertson Stevens and William Merrick, borrowed an antenna design from radio astronomers to deal with this matter. They chose a parabolic dish 85 feet in diameter, equipped with an equatorial polar mount based on astronomical requirements for tracking celestial bodies. The gear system that moved the antenna was polar mounted; that is, the axis of the polar or hour-angle gear was parallel to the polar axis of Earth and thus pointed toward the North Star. This gear swept the antenna in an hour-angle path from one horizon to another. The declination gear wheel, the smaller of the two gears, was mounted on an axis parallel to Earth's equator, thus allowing the dish to pivot up and down. The gears could be moved either separately or simultaneously to provide precise tracking. The equatorial mounts used for the deep space dishes were better suited for tracking interplanetary spacecraft; they allowed principal movement around only one axis, since the rotation of Earth provided the other.

The standard ground station antenna was a large parabolic reflector—a perforated metal mirror that looked like an inverted umbrella and was usually called a "dish." The antenna and its supporting structure stood 10 to 20 stories high and weighed hundreds of thousands of pounds. Since the antenna had to point directly at the object being tracked to receive the strongest signal, a servo system normally operating in a feedback or slave mode was used. Pointing angle information was based on trajectory data predicted by computer in advance and then updated by actual trajectory data obtained during a mission. All parts of these antennas were so precisely balanced and aligned that, in spite of their weight, they could be rotated very sensitively, with only small deflections or vibrations that might cause the signal to be fuzzy.

Astronomical antennas that were the starting point for deep space tracking and data acquisition antennas did not have two-way communication capability, for there was little reason to broadcast commands to the stars. Thus it was necessary to provide the communication transmitters and the feeds that would allow the dishes to serve as transmitting antennas as well as receivers. This called for diplexers to permit simultaneous transmission and reception using a single antenna. Added capabilities were referred to as the uplink and downlink functions. Devices were also added to the antennas for tracking the spacecraft of interest and for "closing the loop" in the sense of driving the antenna-pointing mechanisms.

Tracking requires two parameters: (1) a measure of angular displacement for the spacecraft with respect to a reference system on Earth and (2) measurement of the distance from the tracking antenna to the spacecraft. The angular measurements can be obtained by accurately calibrating the directional pointing system for the antenna as a function of its drive actuators. The distance measurement is based on the Doppler principle, well known for its use in determining the relative speed of a celestial body or a star with respect to Earth. The so-called Doppler shift is really the apparent change in frequency of a signal reflected from or emitted by a moving object as the object moves toward or away from the observer. Everyone has experienced the Doppler effect: the whistle of an approaching train sounds high pitched, and the pitch drops as the train passes. The same thing happens to radio signals, and it is possible to accurately determine rate of change in distance by measuring the frequency shift.

Early spacecraft used one-way Doppler; that is, signals from the spacecraft were transmitted to the ground and changes in frequency were measured in the same way the sound from a train whistle might be measured for its change in frequency. This technique depended on knowledge of the precise transmitting frequency of the spacecraft; its accuracy was limited because frequencies were always subject to change. Two-way Doppler was developed to increase accuracy from about 90 feet per second to as little as 1 inch per second. The concept of two-way Doppler is simple: a precise signal transmitted from the ground is received by the spacecraft transponder and retransmitted at a new frequency in a precisely known ratio to the one received. This allows measurements of frequency change in the signal on the way up and on the way down, tremendously increasing the precision of the Doppler information and the velocity calculations. Using two-way Doppler, the distance to a spacecraft several million miles away could be determined within 20 to 50 statute miles. Later, an automatic coded signal in conjunction with the Doppler information provided measurements with an accuracy better than 45 feet at planetary ranges.

Because the Doppler shifts due to changes in the velocities of spacecraft varied widely, receivers had to be continually tuned to a narrow range of frequencies. This was a troublesome problem until a technique was found that provided a phase-lock method of signal detection, maintaining an automatic frequency control and keeping the receiver locked with the received frequency. Thus, even though the frequencies were changing with the speed of the spacecraft and the relative speed due to the rotation of Earth, it was possible to maintain a coherent tracking of the spacecraft under widely varying conditions.

Automatic phase control origins date back to the 1920s and 1930s, but the first known application was in a horizontal line synchronization device for television in the 1940s. Rechtin and R. M. Jaffee showed in 1955 how a second-order phase-locked loop could be used as a tracking filter for a missile beacon, and specified how to cope with Doppler signal shifts in weak noise. W. K. Victor further developed the theory and practice for automatic gain control for such closed loops in conjunction with his many other contributions to spacecraft tracking.

According to senior JPL engineers, the transition from vacuum tubes to solid-state technology was not without trauma. It is understandable that a change from such a highly developed and known technology to the mysterious new promise of transistors and diodes caused project engineers many headaches. Most engineers involved with spacecraft hardware were familiar with the shortcomings of vacuum tube technology; vacuum tubes were particularly subject to problems caused by the severe acceleration and vibration environment during rocket launch. However, tradeoffs involved in dealing with known qualities versus the uncertain effects of a new technology were difficult to assess. It took many years to develop confidence in the application of solid-state electronics, even though the principles were proven and understood. Ranger and Mariner spacecraft were among the first to be fully committed to the use of such devices, with the major exception that their power amplifiers were vacuum tube triodes.

Robertson Stevens cited three major factors responsible for the low bit rate that was achieved with Rangers and early Mariners: limitations in power, limitations in antenna size, and low transmitting frequencies. One of the reasons for power limitations in early missions was the fact that transmitters were powered by vacuum tube triode amplifiers which were heavy and inefficient power consumers. It was not until traveling wave tube amplifiers came into use (the first lunar and planetary spacecraft application was Surveyor) that a significant increase to 20 watts was made in transmitter power.

The antenna size was of course limited by the difficulty in packaging antennas to fit within the shrouds on top of boosters, as well as the weight available for such structures.

The frequency limitation was related to several factors, not the least of which was the greater accuracy of antenna geometry required for operation at high frequencies. In addition, there were problems in discriminating and dealing with high-frequency signals. Political factors also influenced the use of new radio frequencies; there was considerable international concern over the allocation of the usable radio frequency spectrum.

The early Ranger and Mariner missions operated at L-band frequencies of about 890 to 960 megahertz. During the middle of the Ranger activity, there was such demand for aircraft communications at these frequencies that a changeover was planned to S-band (2110 to 2300 megahertz), making the L-band region of the spectrum available for Earth communication links with aircraft and other users. This change was to have occurred with Ranger 10 and subsequent Rangers which were canceled; the upgrade in frequency was made in 1964 for Mariners 3 and 4.

The conversion to higher frequencies required a major modification of equipment and procedures at all DSN stations; however, this change had merit for space application once the engineering had been done. The simple matter is that, for an antenna of a given size, higher frequencies allow narrower beams, higher gains, and improved performance. The early Explorers used signals in the 100-megahertz region, and the antennas spread data in all directions; the current capability of Voyager at frequencies of 8500 megahertz (almost a hundredfold increase) provides energy 105 times more focused because of the narrow beam. Of course this translates into a burden for accurate attitude orientation or pointing, both for the spacecraft and ground-based antennas. As time passed, it was possible to build larger antennas for ground use that had the stability required for high frequencies in addition to greater collecting areas. The first DSN dishes were about 26 meters (85 feet) in diameter; these were later supplanted by 64-meter (210-foot) dishes with greatly increased signal-gathering capability.

At the time the DSN was being initiated, signals returned from space were amplified with tube amplifiers, which were connected by cable from the antenna and, being large, bulky devices, were housed nearby. Because they operated at high temperatures, they added radio noise to the signals received from space. In addition to their own noise, the cabling and mechanical filaments picked up noise from extraneous sources, so that the total signal-to-noise ratio was quite low. Even though inefficient, these amplifiers could amplify the signals as much as 1012 times the received signal strength; this was of course necessary to make the very weak signals useful.

Early in the 1960s parametric amplifiers were developed. These were applications of solid-state technology and used cooled devices operated at temperatures much lower than the hot elements in vacuum tubes. Parametric amplifiers provided something like a factor of 10 improvement in the reduction of noise and therefore greatly increased the amplification processes for weak signals.

An outgrowth of this cooling amplification technology was the development of the maser, an amplifier that used elements cooled by liquid helium to 4 K, very close to absolute zero. Maser is an acronym for "microwave amplification by stimulated emission of radiation." (I was amused when talking with Stevens, who had been involved in the technology before, during, and after the maser was invented, that he was able only with some effort to recall the labeling for each of the letters in the acronym. This is typical of the problems we engineers generate by using the "alphabet soup" approach for describing things.) The heart of the maser amplifier is a synthetic ruby crystal, immersed in liquid helium to keep it at a very low temperature. It operates with a "pumped-in" source of microwave energy to augment the strength of the incoming signal without generating much internal noise.

I was told an interesting account of maser development involving Walter Higa, a JPL engineer who went to Harvard and worked as an apprentice to the inventor of the maser amplifier concept. Higa returned to JPL and immediately went to work to build a maser for space application. It was obvious that to receive the full benefits of such an amplifier, it should be located at the feed of the antenna, as near as possible to the point at which the signal was collected, thus avoiding the addition of noise by cables that might sense spurious signals or other interference. This meant that the liquid helium cooling system also had to be on the antenna and move with it as it tracked a spacecraft. In the very early application, liquid helium was available only in large vacuum Dewars, and the reservoir on the antenna itself had to be refilled about every 10 hours by a man raised with a cherry picker crane device. After doing this onerous chore for some time, an ingenious JPL technician who had been an automotive mechanic developed a refrigerator system that eliminated this unpleasant duty. His scheme involved a small refrigeration unit with connections from the base of the antenna, providing the generation of liquid helium on the antenna from a source on the ground, so that the operation could be self-sustaining.

Regarding the noise contribution of the system, the maser and the large dish technologies have been developed so well that there might not be much more to gain by further refinements. According to Stevens, an improvement of less than 20 K in noise temperature is theoretically possible. Of this amount, about 4 K is attributable to the background noise of space which cannot be eliminated, about 3 to 4 K is due to maser inefficiencies, and about

6 to 8 K is due to atmospheric effects, depending on the frequency used. Of course, the higher the frequency, the better, although weather definitely affects the noise, even at X-band, which is 8500 megahertz. A problem always exists because of the antenna temperature: this is caused by the proximity of Earth and the reflective objects which radiate heat to the antenna. It is estimated that a factor of two in gain might be possible, even if an antenna were located on the back side of the Moon to minimize the heating and noise effects.

One additional trick that is being used to improve capability is called "arraying." This involves the concurrent use of several antennas in the same general region to effectively increase the dish area. By using four antennas in Australia, for example, a data rate of 29.9 kilobits per second can be returned by the Voyager spacecraft when it passes near Uranus in January 1986. And this remarkable rate is achieved using a spacecraft antenna only 3.6 meters in diameter, transmitting signals over a distance of 3 billion kilometers!

An interesting outgrowth of deep space tracking is that the known location of stations on Earth was improved greatly in the process. As a result of the Mariner mission to Mars in 1964, it was estimated that the absolute location of the Goldstone tracking station was improved from an approximate position within 100 meters to within 20 meters. This figure has been improved during subsequent missions to within less than 1 meter.

The way in which station location is determined from the Doppler data may be understood by supposing the spacecraft to be fixed in space with respect to the center of Earth. The only Doppler tone would be caused by the station's rotational velocity along the direction to the spacecraft: therefore, the observed Doppler tone at the station depends on the latitude, longitude, and radius from the center of Earth. Since thousands of measurements were obtained during the many tracking passes of the network stations, it was possible to deduce the proper combination of station location errors to match the data. It is also interesting to note that the masses of the Moon and the planets were determined in a similar fashion. In the case of the Moon, for example, the variation in Doppler tone was due to the movement of Earth around the Earth-Moon system's center of mass, or barycenter. Earth makes one rotation around this barycenter every 28 days at a speed of 27 miles per hour. This could be measured accurately by the tracking system.

In every case, the orbit of a spacecraft flying past a planetary body is deflected by the gravity of that body. The amount of deflection, coupled with the knowledge of the distance from the center of mass, allows scientists to calculate very accurately the mass of the body in question. Tracking data obtained from Lunar Orbiter spacecraft produced data which allowed the scientific discovery of mass variations within the body of the Moon. Paul M. Muller and William L. Sjogren were able to use the accurately determined variations in the track of Orbiter around the Moon to identify mass deviations and even to locate them approximately beneath the surface of the Moon. These anomalous concentrations, called "mascons," were discovered to be present in the great circular mare basins, suggesting that large chunks of heavy material may have sunk into a plastic, perhaps molten Moon until the gravity field was restored to equilibrium. The findings were not only of interest scientifically, they were also significant for planning Apollo missions to the Moon, because the mascons definitely affect the orbital and trajectory parameters of lunar spacecraft.

The radio signals used for tracking purposes have also served a number of additional scientific studies. Whenever a spacecraft flew past a planet in a way that caused the radio signals to pass back through its atmosphere, the attenuation and distortion of the signals allowed a great deal of deduction about the nature of the planet's atmosphere and ionosphere. Such experiments gave the first definitive information about the atmospheres of Mars and Venus.

Although direct communication links with spacecraft were prime considerations, it must be remembered that a large, Earth-based complex was involved in the total process. Included were the Space Flight Operations Center colocated with the Space Flight Operations Facility, the Launch Control Center colocated with the launch facilities at Cape Kennedy, certain Atlantic Missile Range stations, and an interconnecting ground network of radio and telephone systems. In many cases, getting data back to JPL after its receipt at a Deep Space Station presented significant challenges. Problems often developed with leased landlines or transoceanic communications—problems that were made more difficult because of the coordination involved. The curious anomaly of being able to communicate millions of miles between planets with greater assurance than from points on the surface of Earth always puzzled me.

In recalling mission activities during years of association with lunar and planetary programs, it seems to me that the telecommunication systems probably were the most dependable of all. I know of no major difficulties resulting from technological mishaps or from overestimating the capability of the tracking, data acquisition, and command process. In discussing this subject with a respected JPL project engineer, he offered the opinion that the telecommunications guys always cheated in the game of balanced design margins for spacecraft. He said they made a practice of computing margins based on the simple addition of all factors and were never forced to use the statistical probabilities that most other engineering tradeoffs involved. As a result, he thinks they normally enjoyed greater margins and were able to do more than was predicted. If he is right, this practice may have resulted in some unfavorable design compromises in other areas; however, it always made *me* feel good knowing that we could count on telecommunications operations to produce the promised performance.

The Matched Pairs—Viking Orbiters and Landers

Although not given much thought at the time, our opportunity to consider different ways of exploring a planet and to plan to do so from scratch was unique. Until our generation, the only planet men had been able to explore was Earth, and its exploration had begun from a local point of view, by walking in ever-widening circles. After a time, men viewed valleys from mountaintops and finally saw both mountains and valleys at the same time from aircraft, but never from the perspective that would have been afforded space travelers arriving from elsewhere in our solar system.

After several years of maturing experiences, I am amused to recall our blasé approach to planning the first missions to the Moon and planets. As engineers and scientists, we had confidence in our abilities and the technologies available, and we simply set about planning to do the things that needed to happen if we were to achieve our goals. Occasionally it would occur to me that we were very lucky to belong to the first generation having such an opportunity, but these thoughts were always fleeting and replaced quickly by the demands of tasks at hand.

No matter how one addressed the question of exploring a distant planet, the first requirement was to get closer. The flyby mission was enough of a challenge at first, and that mode occupied us fully for a time. The benefits of orbiting to extend the period of observation and to increase coverage were recognized as next-generation extensions of the flyby mode: landings were ultimately needed to assess the nature of the surface and features of the planet.

Making simultaneous observations from orbiting and landing spacecraft was obviously a desirable means of multiplying returns. The combination of synoptic views from orbit and the detailed information obtained at a specific site on the surface would allow broader interpretations of the "global" properties and provide insight for the interpretation of point information.

The Surveyor program was originally conceived as combinations of orbiters and landers, but because the Moon is a relatively "dead" body, there was no great need for orbiting and landing missions to occur at exactly the same time. Conditions on the planet Mars, however, are dynamic, with frequent changes in atmospheric conditions, changing seasons, frequent dust storms, and frost-covered polar caps; thus, obvious benefits could accrue from concurrent observations. There was another powerful reason for the simultaneous launch of orbiter-lander combinations to Mars and Venus. Because of the roughly 2 years between launch opportunities, complete information would be obtained much sooner if orbiters and landers could be dispatched at the same time.

However desirable, the first opportunity to plan for such a concept came with a program named Voyager. Early studies for Voyager began in 1962; however, it was not until the 1965-66 period that program approval was obtained. The Voyager project plan envisioned the development and use of an orbiter-lander spacecraft combination having broad capabilities for conducting missions to Mars and Venus at several opportunities. The long-range objective was to allow systematic exploration of these planets with two launches per opportunity, using production-like hardware in a manner similar to that being applied to missions to the Moon. Initially the Voyager launch vehicle was to be a Saturn 1b integrated with the Centaur upper stage, a development thought to be relatively straightforward, since both vehicles already existed and were seemingly compatible. As no other planned uses for this vehicle combination existed, it would have been dedicated exclusively to Voyager missions. After a while, the undesirable economics of this exclusive use situation contributed to an alternate decision to adapt Voyager to the Saturn 5 vehicle that was already being used for Apollo. It was a larger and more costly vehicle, but by this time it was well along in its development, and being advocated as a production vehicle for long-term use, making it more attractive for the long-term Voyager program than the Saturn 1b/Centaur.

Those of you familiar with the 1980s achievements of the Voyager spacecraft that have successfully flown by the planets Jupiter and Saturn may be wondering how Voyager was transformed from a Venus-Mars program to an outer planets program. Perhaps at this juncture it is well to explain that the original Voyager program was canceled before it really got going, and that the name Voyager was later given to a Jupiter-Saturn flyby program that had been identified for a time as the Mariner-Jupiter-Saturn, or

MJS, project. A name change was ordered by NASA Headquarters to portray the increase in scope over earlier Mariner-class missions, and since Voyager had been dropped some 20 years earlier, it was decided that the name could be used again. In spite of some JPL concerns that the new program might be tainted with the handed-down moniker from an unsuccessful effort, the program turned out well, as described briefly in a subsequent chapter.

In reviewing the setting for Voyager mission planning, it is helpful to recall the impact of Mariner 4 results obtained during the 1965 flyby of Mars. Like its Mariner 2 predecessor, Mariner 4 was the second of a pair of modest spacecraft launched by an Atlas/Agena, with Mariner 3 suffering a fate similar to that of Mariner 1. This time it was the shroud atop the launch vehicle that caused the heartbreaking failure and not the vehicle itself, but the results were the same. Mariner 4 made the long trip to Mars in late 1964 and into 1965, gallantly photographing the planet with a television camera and returning the pictures at the painfully slow rate of 8⅓ bits per second.

A most significant finding from the close-up Mariner 4 pictures—only 21 and a fraction were taken—was the fact that Mars appeared a lot more Moon-like than Earth-like. To the surprise of everyone, including the scientific community, Mars was found to be heavily cratered, with no evidence in any of the photos of the canal-like features that had been envisioned from astronomical observations using telescopes.

After this revelation by Mariner 4, reasons for the large number of craters were readily forthcoming, yet searches of the scientific literature revealed only a few brief inferences by scientists that such might be expected. During the planning for Mariner 4, I had never heard any suggestion from our scientific advisors that they expected Mars to be covered with craters. This incident made me a little more wary of the profound projections some scientists were prone to make; several of the investigators' reports gave the impression that they, too, were somewhat humbled by the oversight.

The dashed hopes of finding "little green men" was devastating to the support for Mars exploration—especially from administrators and members of Congress. While Mariner 4 results were also disappointing to those directly involved in the "business," some of us felt that the evidence was so scant that we surely ought to conduct a more thorough search before writing Mars off as a dull, lifeless planet. Accordingly, our determined pursuit for approval of more Mariner missions continued while planning began for the Voyager-type program.

Three burning questions were formulated concerning the matter of life on the planet:

> Is there life of some form on Mars now?
> If not now, has there been in the past?
> Could the planet become habitable for life as we know it?

In addition to these life science questions, there were many valid scientific questions about the planet that had not been answered by the Mariner 4 findings—details concerning its body properties, its atmosphere, its mysterious polar caps, whether volcanic activities existed, and the like. As discussions about Mars were again stimulated, a new wave of excitement arose in the scientific community which also began to infect others who had lost interest in Mars after Mariner 4.

Voyager flights were not planned to begin until the early 1970s, so we boldly continued to work toward Mariner Mars missions in 1969 and 1971. The 1969 flights were to be more sophisticated flybys, and in 1971 we hoped to orbit the "red planet," producing maps and other data that would be useful for planning Voyager missions. As it turned out, these Mariner missions were extremely important, and a slight digression is warranted to explain why.

From my viewpoint, the evolutionary advances in mission capability afforded by the smaller Mariner-class spacecraft were more logical steps than the "order of magnitude," scaled-up efforts required to develop and operate Saturn-launched Voyager spacecraft. I was worried that we would find ourselves with all our eggs in one basket—with higher risks and with financial, management, and organizational challenges much harder to control. As a matter of fact, technological improvements in the Mariners had already made them appear capable of addressing the most immediate scientific questions, at least until after we had been able to conduct orbiting missions to observe and map most of the planet.

I made my views known, but the enthusiasm of senior NASA officials for proceeding to large spacecraft and Saturn-class vehicles overswept my more modest ambitions, and I found myself spending more and more time organizing the large-scale Voyager effort. From the outset it was obvious that the program would require a coordinated effort of several field centers. At this time JPL was very busy with the Ranger, Surveyor, and Mariner programs; furthermore, for Voyager we were talking about launchings of multiple spacecraft on a production basis. For a program of this scale, it was decided that the effort should be organized along the lines of Apollo, with a Headquarters program office and a NASA project manager reporting directly to a Headquarters program director. This decision disappointed JPL because it meant that their organization would not be eligible for the same type of lead role they had performed for Mariners, Rangers, and Surveyors. They clearly were to be involved as principals—no other center had as direct experience in planetary programs as JPL, plus direct responsibility for many of the key facilities. However significant the JPL role might have been, the new management concept for Voyager precluded JPL's being given project management responsibility.

The way the program office was finally established, my title was changed from Director of Lunar and Planetary Programs to Director of Voyager and Acting Director of Lunar and Planetary Programs. Donald P. Hearth, who had been Chief of Supporting Research and Technology for Lunar and Planetary Programs, was named Acting Project Manager and quickly began setting up a project office using a cadre of experienced JPL engineers and scientists. A floor of a new bank building was rented in downtown Pasadena, as JPL did not have space that could be dedicated to a new organization of the size required for Voyager. Don began to spend most of his time in Pasadena, but he "commuted" from Washington and never made a permanent move.

Although I became engrossed in Voyager, I was very pleased to be allowed to keep responsibility for directing all other lunar and planetary programs. I was in the thick of things with the ongoing Lunar Orbiter, Surveyor, Mariner, Pioneer, Apollo Science, and related activities, and it would have been a major blow to give those up after so much had been put into them. Had I been required to make a choice between continuing to direct those programs or Voyager, I would have opted to stay with the several smaller programs, even though being Director of Voyager was more prestigious.

The initial plan to use the Saturn 1b/Centaur vehicle for Voyager called for two launches at each opportunity of a 2000-pound orbiter and a 2200-pound landing capsule combination. The 1965 decision to change to the Saturn 5 meant that one launch could be made at each opportunity, with a single Saturn 5 carrying two Voyager orbiter-lander combinations having a gross weight of 62 700 pounds! Stacked on one Saturn would have been two orbiters, two landers, two surface science laboratories, and all the attendant retro motors, entry capsules, and ancilliaries. Such a mission would have truly been an exploratory expedition, launched simultaneously on a single rocket. This would indeed have put a lot of eggs in one basket!

In beginning the Voyager program, Headquarters had to give the major field centers their project assignments. To ensure that proper attention was paid to this matter, I organized a Voyager Board of Directors with the blessings of Ed Cortright and Homer Newell. A general management plan had been roughed out, prescribing program direction by our management team at Headquarters and in Pasadena, with major center participation by JPL, Langley, Marshall, and Kennedy. The directors of these centers, William Pickering, Floyd Thompson, Wernher von Braun, and Kurt Debus, were very cooperative in agreeing to serve on the Voyager Board with Don Hearth and me.

Our understanding was that we would meet quarterly to establish organizational relationships and to develop guidelines for all major activities. I knew that such a beginning relationship with the center directors meant we would get the right kind of people assigned. With regularly scheduled quarterly meetings, we would also have a good means of reviewing progress and, if necessary, dealing with problems. The first Voyager directors meeting was held at NASA Headquarters on April 27, 1967, and the second was 3 months later at JPL. The board started off well, and it looked as if the Voyager program had everything going for it shortly after it was officially initiated.

A Lunar and Planetary Missions Board was also established at about the same time through the National Academy of Sciences to provide advice concerning the science activities related to the Voyager missions. The group was chaired by Harry Hess, a renowned geologist from Princeton, and included a "Who's Who" list of scientists from astronomy, life sciences, geology, radio astronomy, and biology from around the country.

Unfortunately, all this administrative and scientific support for the program was not enough. In the summer of 1967, shortly after the management and planning efforts were established, the Voyager program was dealt a death blow when Congress pared it completely from the NASA budget. The problem was not so much sentiment against Voyager per se as a generally perceived need to stop what some considered a runaway budget situation, making this large new program a target for a major reduction. Everyone involved fought to save the program, but by September it became clear that appropriations would not be forthcoming to sustain the momentum of

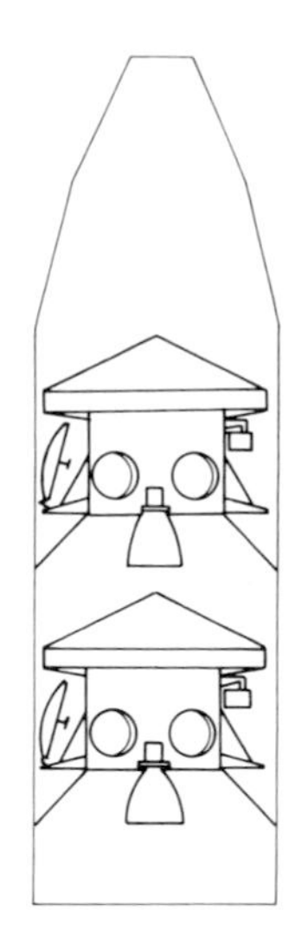

Characteristic	Baseline	Typical growth potential	
	1973	1975	1977/1979
Spacecraft bus/orbiter	2,500	2,500	2,500
Capsule	5,000	6,000	7,000
Propulsion	13,000	14,000	15,000
Total (one planetary vehicle)	20,500	22,500	24,500
Net injection weight (two planetary vehicles)	41,000	45,000	49,000
Shroud/adapter	9,300	9,300	9,300
Project contingency	5,000	3,700	2,700
Gross injected weight	55,300	58,000	61,000

Original Voyager concept

Voyager. The political climate was such that support simply could not be mustered for the combined requirements of Apollo and a large planetary program. Continuing to fight for Voyager would clearly have compromised our support for other important commitments, with no guarantee of success. We bit the bullet, closed the project office at JPL, and disbanded the Board of Directors.

Our struggle to maintain continuity for planetary exploration through Mariner-class missions succeeded, with only one failure in the remainder of the series. Mariner 5 successfully performed a flyby mission to Venus in 1967, Mariners 6 and 7 made flybys of Mars in 1969, Mariner 9 orbited Mars in 1971, and Mariner 10 flew by both Venus and Mercury in 1974. These Mariners effectively bridged the gap that would have developed if they had been abandoned in favor of the Voyager effort. It is my view that the total planetary program turned out well, taking into account the fact that the Mariners and the Viking replacement for Voyager were complementary, affording the scientific community a meaningful basis for continuing study through several planetary opportunities.

Mariners 6 and 7, both successful flybys, clearly showed their superior technology over Mariner 4, although their design was begun with the ground rule that the spacecraft would be the same and only the scientific instruments upgraded. Unfortunately (or fortunately), the best way to upgrade the science return required technological advances in the spacecraft. The most notable improvement was in the communications bit rate—increased from the paltry 8⅓ bits per second of Mariner 4 to 16 200 bits per second. In addition, the data subsystem was upgraded markedly with two specially designed tape recorders to meet requirements 35 times as great as those of Mariner 4. Improvements were also made in the telemetry subsystem: the remarkable scan platform had the ability to adapt to changing requirements as well as to accommodate modifications in the second flyby mission, thus allowing the two spacecraft to perform complementary rather than repetitive roles.

The greatly improved images of Mars obtained by Mariners 6 and 7 provided many surprises. Our concept of Mars changed again, from a barren Moon-like planet that appeared lifeless to a more Earth-like body having many types of terrain, clouds, variations in atmosphere, and evidence of erosion, strongly suggesting that water had once been abundant. These findings led to a revitalization of interest in Mars as a place where life had been harbored at some time, if not in the present. This interest was quickly shared by scientists, with administrators and politicians becoming advocates as well. The result was continuing support for the orbiting missions of Mariners 8 and 9 planned for the 1971 opportunity. If the planet could not be surveyed concurrently from orbit and on the surface, at least the next most vital steps, conducting orbital surveys, preparing maps, and allowing more sophisticated planning for landing site selection, could proceed.

After a string of successes, Mariner 8 became just another statistic as a result of launch vehicle failure. The Atlas performed well, and powered flight proceeded normally until shortly after separation and ignition of the Centaur stage. At that point, a pitch control problem in the Centaur flight control system allowed the stage to tumble and shut down. This disheartening loss was followed by the usual reviews, modifications, and adjustments, but these were completed in time for Mariner 9 to be launched successfully.

In keeping with the general goals for planetary exploration, Mariners 8 and 9 were to provide detailed photographic surveys of the planet at much higher resolution than ever before. Special studies were to be made of the so-called "wave of darkening" along the edges of the polar caps, including measurements related to temperatures, surface composition, the presence of water molecules, and the existence of other conditions generally relevant to the question of life.

In late September 1971, astronomers who were keeping a watch on the planet saw a bright yellow cloud forming in the southern region known as Noachis. Dust storms had been seen on Mars before, but this one was of special interest, as Mariner 9 was to experience it from nearby. When this storm was in its fifth week, it peaked—apparently worse than any that had ever been observed both in area and duration. By November 14, when the spacecraft was placed into orbit by firing its retro rocket, the worst of the storm had subsided, but only five distinct surface features could be identified. Conditions definitely were not those envisioned when the mission was planned, so plans had to be changed.

Fortunately, Mariner had the capability for reprogramming, a highly desirable feature for an exploratory mission, made possible by improvements in technology. Actually, observing the changes as conditions improved provided much new insight, and since Mariner 9 was viable for more than a year (the design goal was 3 months), a thorough study of this Martian dust storm was possible.

One of Mariner 9's revelations was a giant volcanic mountain, named Olympus Mons, and an almost unbelievable canyon system, far larger than Earth's Grand Canyon, named Valles Marineris in honor of its Mariner discoverer. Of course, the multiple-orbit imaging coverage provided by the long-term mission allowed cartographers to prepare detailed maps of Mars and provided scientists with several types of data for speculation about conditions on the surface.

Not long after the termination of the Voyager project, a new landing mission concept was born from the ashes. Advanced technology work had been continuing for several years on capsules designed to survive a hard landing; results were encouraging to those who hoped to obtain important data from the surface of Mars. In addition to the scientific stimulus, there had always been broad support for landing on Mars; this was, after all, a clear milestone in the space race that the Russians had been trying to achieve for a long time, if only for its propaganda value.

The new mission required a new name to give it a fresh start and to distinguish it from Voyager. Viking was the name chosen, and the first flights were proposed for the launch opportunity in 1973. The Viking program, proposed to be a bargain at only $364 million, was initially conceived to involve a Mariner-derived orbiter and a simple, hard-lander spacecraft. Congress approved the project in 1968, but it soon became apparent that funding and the scope of the mission did not mesh. After the grandiose studies and planning that had been done toward Voyager, we experienced difficulty in scaling down. Matters were made worse by a strong desire to make a quantitative advance beyond the 1971 orbiting missions, requiring the landing spacecraft to have a greater capability than a hard-landing capsule. After a good bit of trauma and failure to match requirements and available funding, launch was postponed from 1973 to 1975, with a continuing program for development throughout the 2-year interval. Many of us were disappointed to pass the 1973 Mars opportunity, for the planetary orbit geometry of Mars and Earth at the time would have allowed the largest payload for a given sized launch vehicle for years to come. But as it turned out, the Viking missions in 1975-76 were probably better for the delay.

The postponement was considered to be a mixed blessing. The total project cost had to go up because people were kept at work longer; the delay provided opportunity for better planning and application of new technologies. Different management arrangements were worked out, borrowing from earlier project experiences and from the concept established for managing Voyager. At this time it was agreed that project management for Viking would reside in the field. Although I had accepted the plan for Voyager and had been named to direct the program from Headquarters, I never did think this was as sound in concept as making a field center responsible. The reason was simple: a manager needed a qualified staff at his fingertips to deal with management problems, and I did not think that we could ever assemble such a team at Headquarters. The Apollo program had been managed that way, but the Apollo Headquarters management team depended on Bellcomm, Inc., a complete systems organization under contract, which we could not have for Viking. Although Apollo was a successful program, I was never convinced that it could not have been managed successfully by a field center along the lines employed for lunar and planetary programs.

Viking missions were based on using the Titan 3C/Centaur launch vehicle instead of the Saturn, so direct involvement of Marshall Space Flight Center was no longer required. Lewis was responsible for Centaur development and for integration of the Titan 3C with the Centaur. Thus, Lewis was the obvious choice to manage the launch vehicle system. Either JPL or Langley might have been chosen for the project management center assignment, but three factors favored Langley: (1) Langley was truly a NASA center and not a "contractor" operation that, at the time, was somewhat out of sorts with NASA Headquarters, (2) Langley had successfully completed the Lunar Orbiter project and had a ready team with no other project assignment at the time, and (3) Langley had a strong research capability to back up development of a new landing vehicle. The landing craft was to be built by a contractor, but the engineering and testing of landing systems, including entry aeroshell, parachute, and landing gear, nicely fitted the Langley background.

JPL was the obvious choice to manage the Viking orbiter system as well as the Deep Space Network. The Mariner-derived orbiter was to be an upgraded adaptation of a Mariner bus designed to transport the lander to Mars and provide injection into orbit. JPL would also play a vital role in space flight operations, as they were responsible for the Space Flight Operations Facility, where mission operations were conducted. However disappointed some JPL members may have been, there was no evidence of bitterness as they turned to the tasks assigned and performed them admirably.

By the time Viking began, there had been enough launches of satellites and interplanetary spacecraft that many people tended to think of Viking as just another, slightly more sophisticated mission, using existing technologies. Actually, this was not true, for the Viking project elements, including both hardware and software, were an order of magnitude more complex than anything that had gone before.

Perhaps the simplest way to explain this premise is to describe the Viking spacecraft just prior to launch. A launch vehicle manager might, out of habit, refer to it simply as the "payload" awaiting launch atop his rocket, but it was really a combination of four spacecraft, each with a different function and purpose. Completely separate yet tightly integrated entities were an interplanetary bus, an orbiter, an entry capsule, and a lander.

For transporting instruments and equipment to the vicinity of the planet there was the "bus," an interplanetary vehicle with attitude control, thermal control, power supply, communications link, midcourse correction capability, and all the systems required for a Mariner flyby mission. To perform the retro maneuver at the planet a relatively large rocket motor was required that could survive the long transit period of the transfer orbit and then be controlled precisely to inject the spacecraft into a preselected orbit.

After serving as an interplanetary spacecraft and injecting into Mars orbit, an additional duty of the bus, now an orbiter, was to serve as a launch platform for the entry capsule. This required precise attitude orientation, timing, and separation signals for ejecting the capsule so that it would enter the atmosphere and descend toward the surface. After this, the orbiter would observe Mars from orbit, in much the same way that an orbiting Earth resources satellite might observe our planet. One continuous and very important supporting function for the orbiter throughout the mission was its role as a communications relay satellite for transmitting data from landers to Earth.

In the early days of missile development, learning to design and build vehicles capable of reentering Earth's atmosphere had been a significant technical challenge. To do the same thing for another planet with an atmosphere far less defined than that of Earth was the challenge faced by Viking entry capsule designers. Apollo and Viking had generally similar design requirements for an aerodynamically stabilizing shape, thermal protection from heating, and base structure and retro motor integration; however, the Viking entry capsule also had to deploy a landing spacecraft at the proper time without damaging its complex equipment and appendages.

A specially developed parachute system was carried to slow the descent for landing; after decelerating the entry capsule to about twice the speed of sound, further use of atmospheric drag was thus made. The parachute system demanded technological developments beyond those being used to return sounding rocket payloads to Earth because of the different atmosphere and approach conditions on Mars. The last official duty of the parachute system was to pull the aeroshell base structure away from the lander spacecraft so that it could extend its landing gear and prepare to land.

In addition to its engineering tasks, the entry spacecraft provided for scientific measurements during its passage through Mars' tenuous atmosphere. Data were collected and transmitted to Earth; thus, the Viking entry system also provided for in situ examination of the unknown Mars atmosphere. This alone was the equivalent of a sounding rocket mission into Earth's atmosphere.

Because attention was focused on the activities of the orbiter and lander spacecraft, the achievements of the entry spacecraft, its parachute, and complex systems were largely unheralded; a few years earlier, these would have been regarded as very significant. Of course, had any components of the entry systems failed to work, their importance to the success of the entire mission would have been painfully obvious.

Most people would recognize the lander spacecraft as a major design challenge, although by the time Viking was being designed the Surveyors had removed some of the doubts about the technical feasibility of developing such spacecraft. Nonetheless, designing landing spacecraft for Mars very nearly required starting from scratch. A major factor was Mars' atmosphere, for during the landing and touchdown phase, aerodynamics had to be considered for stability and control as well as rocket performance. This was not a requirement for landing spacecraft on the airless Moon. Mars also has a significantly greater gravitational pull to overcome than does the Moon, so all-new requirements existed for engineering design. And, even though the attitude stabilization, Doppler radar, and retro rocket systems were generically similar to those of Surveyor, they were different enough to demand special detail in their design. An entire book could be written about the Viking landers and their almost unbelievable qualities; to discuss them in a few paragraphs is almost an injustice to those who should be credited for the design, development, and operation of these magnificent, self-sufficient, automatically controlled yet responsive machines.

The first duty of the spacecraft was to automatically land safely on the unknown surface of Mars without damaging any of the precious cargo of scientific instruments. To do this it had to determine how far above the surface it was, adjust descent and lateral velocities so as to touch down within prescribed limits, and then shut off the rocket motors at precisely the right speeds and altitudes. Because of the 20-minute lag in communications between Mars and Earth at the time of landing, the spacecraft's makers on Earth were absolutely no help in performing the real-time activities necessary to successful landings. I clearly recall discussions in the Space Flight Operations Facility at JPL during the period when we knew that either the landing had been done successfully or the lander had crashed, as we anxiously awaited data that would tell us what had happened. In some respects this was like watching a TV replay to learn the outcome of a sporting event that had already been decided.

After landing, the spacecraft became a science laboratory extraordinaire. It was at the same time a weather station, a geophysical observatory, a life sciences chemistry lab, a remote materials manipulator and processor, a data acquisition and processing station, and a data transmitter. It had its own power supply in the form of two radioisotope thermoelectric generators that used plutonium 238 to provide 70 watts of continuous power. It also contained a computer-centered "brain" called a guidance, control, and sequencing computer (GCSC), which could contain up to 60 days of instructions. Of course, the memory could be modified or updated from Earth when changes seemed necessary, but the spacecraft could easily take care of itself during the 12-hour periods when it was out of sight of Earth because of the rotation of Mars.

Because a major goal of Viking missions was the search for life, it was essential that Viking landers not take any form of life to Mars. Thus, the spacecraft had to be sterilized after they were built and tested. To achieve the prescribed degree of sterilization necessary to satisfy internationally established planetary quarantine requirements, lander spacecraft were sealed in their bioshields and baked at temperatures above 113° C for about 24 hours. This had been shown to be adequate to ensure that the chances were less than 1 in 10 000 that a single organism would be transported to Mars from Earth.

Heating for the purpose of killing living organisms also produced risks to lander hardware, especially to electronic components. In order to plan for this last-minute treatment before launch, a great deal of component and materials testing had to be done before selections were made in the design process. Even so, this exposure to unusually high temperatures was made with a certain amount of concern about its effect on the lifetimes of critical components. Many lingering fears remained after the sad experiences with Ranger that were believed related to sterilization requirements.

The science instruments chosen for Viking lander spacecraft were selected very thoughtfully in accordance with major mission priorities and the state of the art in instrumentation technology. Not only was it critical that each instrument be capable of making contributions to knowledge on its own, but most instruments had to become components of a laboratory-like complex. Findings could be expected to be mutually reinforcing, such that the whole would be greater than the sum of the parts. In some cases, a component intended primarily for a scientific purpose also served a supporting function in another scientific investigation.

The choices for meaningful experiments were many; the final complement of Viking instruments was believed to address the highest-priority questions about Mars. There were cameras to see and observe as an inquisitive explorer would have done; meteorology sensors to measure and record the atmospheric conditions and report on the weather; "tools" for scratching the surface and for quantifying the physical properties of the soil; experiments to determine chemical constituents, mineral content, and composition of the soil and atmosphere; and, very importantly, there were three ways of measuring biologic activities that would answer burning questions about life on this neighboring world.

Of all the scientific instruments that have been carried into space, none are more appealing to most of us than cameras. Through our eyes we see things for ourselves; through the cameras onboard Viking our eyes were allowed to sense the mystery and beauty of this distant world as if we were there. The cameras used in the Lunar Orbiter were sometimes referred to as

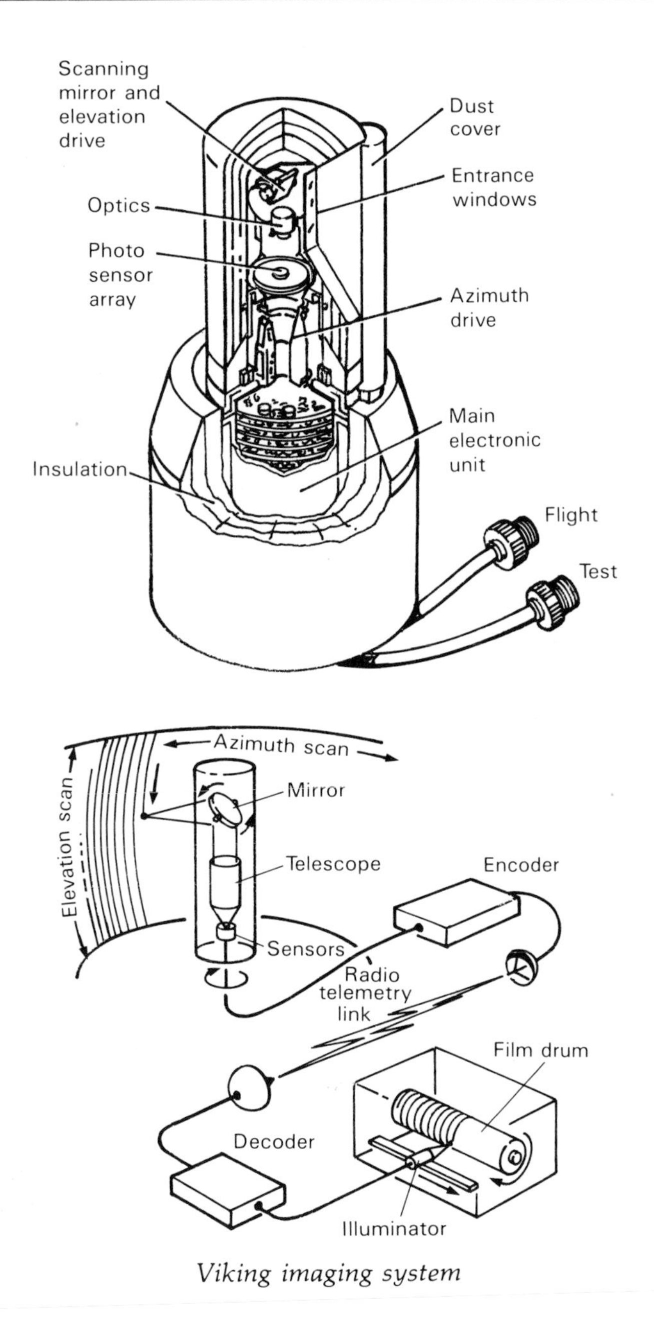

Viking imaging system

"large Brownies," simply because they functioned much like the hand-held cameras seen around us every day. The other cameras commonly used in space were video or television cameras, somewhat less familiar at the time, but now commonly in use. Viking lander cameras differed from both of these in several respects. They employed engineering principles that had been used before, but the method of implementation was different. They were called "facsimile" cameras because of the way they viewed and reconstructed scenes.

In principle, an image can be constructed by a light sensor that sees "elements" one at a time. If the elements are viewed in a row or line, and stored or transmitted electronically, they can be reconstructed as elements having the same intensity. Elements reconstructed and placed in contiguous lines combine to become whole images. Mariner 4's pictures of Mars were comprised of 200 TV lines with 200 picture elements per line (called pixels)—each represented by 64 shades of gray. Over 8½ hours of transmission time was required to return the data for a single picture.

In talking about the process of data return and picture reconstruction, we jokingly discussed the possibility of putting up a large billboard with 40,000 nails on which to hang small square coupons. With a supply of coupons representing the 64 gradations, it would have been possible to hang the numbers in place as they were returned by telemetry so that a picture would be revealed, almost as if painting by number.

A process close to this actually materialized, as the numbers representing shades of gray were printed out sequentially on paper ticker tape. The columns of numbers representing a vertical strip 200 pixels long were then stapled side by side on a piece of beaver board, and colored crayons were used to color corresponding shades of gray. The result was a false-color image showing the edge of the planet in some detail, as well as the varying intensity sky above.

This historic picture was later framed and hung in the JPL Director's office area—a fitting memento of the first successful close-up imaging of the planet Mars. The display is now a museum piece, and destined to be of significant interest to future generations.

The Viking camera made use of light detector, lens, and mirror systems to perform a linescan. A nodding, rotating mirror allowed successive sweeps to reflect an image of the surroundings into the lens. Twelve detectors, three of which had color filters of red, blue, and green, allowed selective images to be recorded and reconstructed in color. By electronically recording the varying intensities of reflection, linescans were converted to digital signals that could be transmitted directly or stored in memory. With simple indexing of position and movement, contiguously placed reconstructions of each line became an image or "picture" fashioned from the composite bits of data.

For the facsimile principle to work, the features being imaged had to remain stationary long enough for the scanning process to be completed. Objects moving about as the facsimile scanning process occurred would not have been seen, except perhaps as small disturbances in line elements. Camera developers jokingly suggested that it was possible that a mobile Mars creature might have moved across the field of view of the Viking cameras without detection; however, since there were no signs that this had happened, even those hopeful of such discoveries had to be skeptical. Although the facsimile principle may seem rather simple, the Viking camera represented a significant advance in the state of the art at the time and was by no means a simple instrument.

Those who followed the Viking mission closely will recall that the first pictures released to the press showed Mars as a very red planet. Unfortunately, this portrayal was generally in keeping with scientific speculation, and 2 days later, when the image data had been thoroughly calibrated, red-faced NASA officials had to tell the press that there had been a slight misrepresentation. For several hours the ground reconstruction process was recalibrated and equipment adjusted; after this was done, some of the redness was reduced. In all fairness, the premature release was probably due to the terrific pressure produced by the desire to share findings with the public as quickly as possible, before completing the data processing checks known to be required.

In a recent discussion with Cal Broome, who had project responsibility for camera development, he indicated that his fondest memories from Viking were of the camera developments and the products they provided. He vividly recalled the experience of viewing the first picture and proudly took the position, "As far as I'm concerned, that's what Mars looked like that day."

He also recalled the trauma that resulted when the all-electronic scanning camera proposed by the Itek Corporation was being considered. A large amount of development effort by Aeronutronics, conducted during the Ranger project, had produced a successful facsimile camera that had been fairly well proven, involving both mechanical systems and electronics in its operation. While it appeared that the Ranger camera would have provided the necessary basic capabilities, it was neither as versatile nor as capable of electronic programming and selective applications as promised for the new concept. In reflecting on the situation, I believe this was simply an example of progress being made so rapidly in fast-moving technologies that excellent concepts became obsolete before they could be used. Regardless, it was the judgment of those responsible for Viking that the advantages of the advanced system outweighed the risks involved in its development, and the Itek concept was selected. This belief was justified by the fact that the electronic scanning cameras worked well. When contact was lost with Viking Lander 1 in 1983, its cameras were still working after 7 years without a glitch.

In his excellent book *The Martian Landscape,* Tim Mutch, Team Leader for the Imaging Science Team, described in a clear and fascinating way the tradeoffs and other aspects of establishing camera design characteristics. When you and I choose a camera from the marketplace, we have no choice but to select from concepts generated by designers who decided what the public would buy. However, in the case of camera design for Viking, the team was able to establish requirements from scratch and iterate them against existing technologies. The most fundamental choice was resolution, the definition of image size for the smallest element to be seen. Selecting the smallest detail to be resolved also implied a maximum field of view, for such was the nature of the tradeoffs. According to Mutch, many of these tradeoffs had been studied for years by Fred Huck, an engineer at the Langley Research Center. With the collaboration of Glenn Taylor, also of Langley, the team was able to examine all the variables of camera performance, including those dictated by spacecraft constraints such as weight, power, and bit rate, and arrive at a balanced design for the hypothesized mission to Mars.

Superficially, the operation of the cameras seemed remarkably simple. The photosensor array and all the electronics that processed the points of incoming light were clustered in a small assembly only 3.4 centimeters (1.3 inches) across. Twelve photodiodes, each able to obtain image data, were mounted so that different focal lengths could be achieved. Some of the photodiodes were equipped with filters of red, blue, and green to permit recreation of color images.

A slot near the top of a small cylinder formed the "pinhole" window through which a small nodding mirror could peer. As the mirror nodded around a horizontal axis, it swept a vertical line, scanning reflected light from the objects in view, while electronic circuits recorded intensities. Five times a second the small cylinder was rotated so that the slot position allowed a new vertical line to be scanned. Indexing for these vertical lines and the timing for the nodding mirror had to be precisely controlled so that each pixel or picture element was contiguous. Actual positions had to be indexed to an accuracy of 0.01 millimeter—about one-tenth the diameter of a human hair—in order for the required resolution to be achieved.

The capability of adjusting signal gains allowed images to be obtained and processed for various light levels, and the variety of photodiodes allowed a selection of amplification for either close-up or distant views. By simultaneously imaging the same scene with two cameras placed about 1 meter apart, stereoscopic views were obtained to permit three-dimensional viewing in the same way that our eyes perform. By any standards, these were remarkable cameras!

In a fashion typical of planetary missions, the launch vehicle used for Viking was specially integrated for this set of missions. The Titan 3 was a military vehicle originally developed by the Martin Marietta corporation for the Air Force. It included a two-stage core rocket system using liquid propellants, plus two large strap-on solid rockets. While not nearly as large as those used to help boost the Space Shuttle, these strap-on solids performed the same function of providing initial acceleration. They were 10 feet in diameter, and each produced about 1.2 million pounds of thrust for about 2 minutes. After burnout, they were jettisoned and dropped into the Atlantic Ocean.

The first stage of the Titan vehicle, also 10 feet in diameter, ignited just before the solids burned out for about 2½ minutes. The second stage then separated and fired for 3½ minutes. Both these core stages used a blend of hydrazine and unsymmetrical dimethyl hydrazine, with nitrogen tetroxide as an oxidizer.

The Centaur upper stage was basically the same General Dynamics-built liquid hydrogen, liquid oxygen rocket used for Surveyor. After separation from Titan, its two Pratt & Whitney RL-10 engines produced a total of about 30 000 pounds of thrust to send the spacecraft on its way. Its relight capability allowed the Vikings to be propelled into a 90-mile-high parking orbit until the right position around Earth was reached for injection into the transfer orbit. The coast periods could vary from 6 to 30 minutes, depending on time of launch. After burnout, the final act of Centaur was to separate itself from the spacecraft and, by expelling its residual propellants, change its trajectory slightly so that it would have no chance of impacting and contaminating Mars. It then became a silent companion to Viking, slowly separating from the spacecraft as both objects coursed around the Sun in the general direction of Mars' orbit.

After the Voyager program was canceled, planning for Viking was begun in a very austere environment. The orbiter-bus was envisioned as a direct outgrowth of the Mariner '71 spacecraft, with a modest scale-up for the additional requirements of Viking. While actually resembling a Mariner and benefiting greatly from its heritage, the Viking orbiter became an entirely different spacecraft. The propellant tanks, for instance, had to be roughly three times the size of those used to provide injection of Mariners 8 and 9 into orbit. The basic structure was enlarged to accommodate the lander aeroshell, and the solar panels were increased in size to provide more power. Over 15 square meters of solar cells supplied about 620 watts of electrical power at Mars, charging two 30-ampere hour nickel-cadmium batteries to be used when the cells were not in direct sunlight or when the spacecraft was oriented for pointing instruments or activating the capsule launch.

Another significant improvement included the addition of extra "brain power" to allow the orbiters to perform more complex functions. Viking orbiters possessed two 4096-word, general-purpose computers that could operate in parallel or tandem modes. These replaced the small special-purpose computers contained in Mariners 8 and 9. The capability for more rapid picture taking allowed for better site surveys and special regional studies. This capability was augmented by tape recorder systems that could store 2.112 megabits per second, with a capacity of 55 TV pictures—over half a billion bits of information.

Viking orbiter communications systems used both S-band and X-band frequencies. A parabolic high-gain antenna, 57.9 inches in diameter, provided for the highly focused transmission and reception of radio energy to and from Earth. This antenna was backed up by a rod-shaped low-gain or omnidirectional antenna similar to the one on Mariner 4, so that no matter what the orientation of the high-gain antenna, communication at a low bit rate was possible. Orbiters also had relay antennas for receiving and transmitting signals to and from the Viking lander spacecraft; this allowed contact between Earth and the landers even when they were on the opposite side of Mars, provided they were in view of Earth.

Transmitter power for the orbiters was about 20 watts, allowing bit rates of 16 000 bits per second. While extremely small compared with the transmitter powers used by broadcast stations on Earth, this was about five times the power Mariner 4 used to provide a bit rate of 8⅓ bits per second. Another significant factor was the development of very sensitive receivers and transmitters in the Deep Space Network, as highlighted by the huge 64-meter (210-foot)-diameter dishes.

While serving as the buses for transporting the landers to Mars, the orbiters had to serve as "hosts," providing the necessary power, thermal environment, engineering status, midcourse corrections, and attitude orientation for capsule ejection. Of course, orbiters continued to serve the landers through the communications relay function after they reached the surface of Mars, but they really performed a major mission function on their own as scientific spacecraft. Their important scientific instruments included two television cameras for conducting site surveys and making maps and topographical studies, an atmospheric water detector, and a thermal mapper to allow studies of temperature variations and to look for hot spots. These alone were adequate justification for the orbiter missions, but in fact, these truly remarkable multipurpose spacecraft did the work of at least four special-purpose spacecraft.

As impressively self-sufficient as the Viking launch vehicles, landers, and orbiters were, three major systems that never left Earth were necessary to their success. These systems formed the connection between the people involved in the missions and the space machines. They were the launch facilities at Kennedy Space Center, the Deep Space Network (based at JPL but spread around the world), and the Space Flight Operations Facility at JPL, where mission operations were conducted.

Visions of the launch complex at Kennedy come to mind immediately; we have all seen television coverage of the gantrys and flame pits in action. Actually many more components—and even complete facilities—were just as vital to the launch operation. Although several were multipurpose, that is, they might also be used for other projects, most had to be especially adapted to Viking requirements. Orbiters were assembled in Building AO and mated with their propulsion systems in the Environmental Safe Facility. Landers were assembled in the Spacecraft Assembly and Encapsulation building, mated with the orbiters, and encapsulated in their heat shields before being moved to the launch pad, where they were mated with the Titan/Centaur launch vehicles. Many of these vital facilities that are sometimes taken for granted did not just happen to be in the right configuration at the right time, but the "heroes" who provided them will never be sufficiently recognized.

The Viking launch vehicle and spacecraft systems presented the most complex array of space hardware ever assembled for unmanned missions to another planet. The combined fleet of interplanetary, orbiting, entry, landing, and laboratory spacecraft that comprised the Viking expedition to Mars in 1975 and 1976 incorporated advanced technologies from almost every major discipline of science and engineering. Dedicated to a single set of goals, most of these automatons were programmed to function effectively with little human intervention; however, all were flexibly reprogrammable to respond to requirements determined by human "masters" on Earth. The Viking team, made up of humans and spacecraft, clearly proved that men and machines can work together in marvelous harmony, provided they are guided by common aims and a willingness to subjugate individual purposes to the greater good.

A Task Force Extraordinaire

The time finally came for the expedition to Mars to begin. In August 1975, almost exactly 10 years after plans for Voyager were initiated, Viking 1 stood tall on Pad 41, ready for launch. The first attempt was scrubbed because of a valve failure in the launch vehicle, and, while awaiting another try, the batteries in the spacecraft discharged. This resulted in the substitution of the second spacecraft while the first was checked to make sure no anomalies had occurred, reversing the planned order of spacecraft launches.

Amid these preparations were periods of anxious waiting for final word on the problem assessment tests. Flashbacks of the long hours of planning, meetings, and frustrations over Voyager occurred, as glimpses of the evolutionary steps toward Viking were recalled. I even had a brief mind-trip back to 1957, when I was an engineer at North American Aviation's Missile Development Division. About the time Sputnik 1 was being launched, we were engaged in a study of a Mars reconnaissance expedition employing ion-propelled spacecraft. Efforts being coordinated by the corporate office involved the Missile Development, Rocketdyne, Atomics International, and Autonetics Divisions. Our ambitious proposal envisioned multiple spacecraft—as many as four—to be launched during the 1964 Mars opportunity, to fly around Mars, gather data, and return to the vicinity of Earth. I led a small team concerned with spacecraft performance, trajectories, and propulsion integration; this study was my first exposure to the excitement of planning a visit to Mars and the beginning of a longing that has yet to be fully satisfied.

After all the years of exposure to missile and space launches, I could not help but think of the launch as the key milestone in any project. Committing to launch meant releasing the precious hardware and all direct control over it to prior judgments; after a missile launch there was nothing one could do to influence the mission, and not much more could be done for the early lunar and planetary spacecraft. For Viking, however, the launch was more like a commencement exercise for a college graduate. It was the end of a long period of programmed experiences and hard work, yet just the beginning of a new and uncertain life. Of course the launch was still critical to success, but by now confidence in launch technologies was high, whereas technologies for landing on another planet were relatively unproven.

For those project members whose special skills had been applied to the engineering, design, and test of the launch vehicle, the trial was over as soon as a good orbit was reported. To be sure, there were data to be analyzed, reports to be written, and post mortems to be conducted, but the victory celebration came first and could be savored without reservation. Soon it would be necessary to concentrate on the tasks for the next vehicle that had gotten behind because the current launch demanded attention, but that was understood and to be expected.

For the spacecraft engineers, the launch had a variety of meanings. In fact, there were so many different possibilities for those who had been involved in spacecraft design, development, and testing that it was not entirely appropriate to think of them as a single class. For the hands-on hardware people, the work was over with the launch, just as it was for the launch vehicle people. Whether the mission succeeded or failed, they had completed the assigned tasks necessary to bring their efforts to conclusion. For some, Viking launches meant the end of a known career; they had been so busy for several years that there had been little place in their minds for thoughts of the future. Suddenly, almost catastrophically, their Viking jobs ended.

For others, the launch meant simply that their jobs would change: some would continue in the same manner, and some were to start a new type of work, with the thread of continuity being provided by intimate knowledge of the hardware or software they had helped to develop. A few were "born operational types" who worked alongside the hardware and software engineers during the development phases, giving counsel, conducting operational studies, and providing planning to support the developments as they went along. For them, Viking really came to life after launch, when, in a real sense, it became a different creature.

An entirely new organization chart was prepared; a number of names reappeared, but there were significant differences. For many who had been involved in the project from the outset, changed assignments meant new titles and work with new groups having different objectives and procedures. Returning to their home center or to JPL after being displaced for weeks or even months at the Cape also meant adjusting to new office environments, as well as to new associates and assignments. Project management officials had wisely begun a transition to the operations phase by reassigning a core group to join the full-time operations experts well before launch. Those already comfortable in this new phase were able to help others adjust to the challenge of conducting flight operations.

A frightening aspect of this, at least to me, was the fact that during the 10 months or so the spacecraft were to be on their way to Mars, a sobering amount of work had to be done in order for the missions to be performed as planned. There had not been time or manpower to "engineer" all the planetary operations until after the launches. The systems had been designed and built with the flexibilities and programming capabilities to allow en route preparation for planetary operations; we would not know how well that goal had been achieved until after the cut-and-try process of simulations or actual operation. Needless to say, the discovery of design deficiencies after the hardware was millions of miles away would not have been very satisfying.

Although there had been a certain amount of new activities and operational training during the cruise periods of the Mariners, their limited capabilities left far fewer options after launch. Furthermore, there were never to be more than two machines operating in the vicinity of Mars at the same time, even if both were totally successful. With both sets of Viking spacecraft on the way, we could look forward to juggling people and facilities to accommodate arrival times, orbital injections, site selections, deboost maneuvers, communications relay periods, and critical orbiter and lander experiment timelines for four very sophisticated spacecraft, all arriving at Mars and requiring careful attention within a few days.

By the time of the launches, Viking's primary missions had been defined and basic arguments settled concerning the scientific objectives and the manner in which the spacecraft would perform. These objectives had evolved in concert with hardware development; this moderated original desires to coincide with hardware and software capabilities. To ensure that the teams and individuals involved all used the same list, the Project Office very plainly spelled out scientific objectives and mission strategies for operational use. There was a general set of these for the two orbiters and a set for both landers. Later, more specific tasks were to be assigned to each, but the pairs of spacecraft were designed to be interchangeable and thus shared the same broad guidelines for mission objectives.

The primary purpose of the orbiters was to obtain pictures, surface temperatures, and water vapor readings. While these data obviously had scientific value in their own right, all were to be used in selecting landing sites for the Viking landers, as the first requirement of every Viking element was to help ensure that landers had the best possible chance of landing safely and conducting worthwhile experiments. The orbiters' scientific instruments therefore had to serve the good of the entire mission before they were to concentrate on important scientific functions. We have already mentioned how the orbiters served as buses for interplanetary transportation, as launching platforms, and as communications relay links.

The second objective for orbiters was to continue the photographic surveys begun by Mariner 9, repeating some coverage of the planet and adding thermal and water vapor measurements during the lifetimes of the landers. Orbital coverage of landing sites and similar areas would be used to extend the meaningful coverage of local lander data, adding emphasis to landing site studies.

The third objective for the orbiters was prescribed with the future in mind, for it specified obtaining images and thermal and water vapor data to help planners in the site selection process for subsequent missions. No one supposed that the first landers would do the complete job of exploring Mars with just two landings, particularly considering the importance of choosing the most hospitable sites for the first landings rather than the most hospitable sites for life to exist. Can you imagine how incomplete your impression of Earth would be if you could observe it only from two flat, smooth spots that had been chosen because they looked like safe landing fields?

The fourth objective specified clearly that orbiters were to obtain images and thermal and water vapor information to be used in the study of the dynamic and physical characteristics of the planet and its atmosphere. At last, thought some scientists, science for the sake of science! Although much data would be gathered in regard to the first three objectives, not until the chores were done in behalf of the whole expedition would priorities rest with the scientists. The objectives list was a reminder of an everyday rule of life: we must ensure our survival before we can achieve higher goals.

There was also a very important fifth objective for orbiters; it called for scientific investigations using radio system data. We think of radios in conjunction with communications, but because the electromagnetic energy transmitted at various wavelengths is affected by the media through which it passes, measuring and analyzing these effects on radio signals as the spacecraft passed behind Mars also allowed scientists to make many deductions. Earlier flyby experiments had generated respect for this "by-product" application of the radio signals for studies of the ionosphere, the atmosphere, and the interplanetary constituents. Viking's use of two frequencies to transmit signals from Mars to Earth would provide a measure of the electron concentration in interplanetary space, enhancing our understanding of communications capabilities for future systems. Radio tracking of the spacecraft transponder during approach and orbit would produce data for calculations of the orbit and mass of Mars, to be deduced from the gravitational influences of Mars on spacecraft trajectories. Finally, the analysis of radio signals as the spacecraft and Mars orbited the Sun would provide information to verify Einstein's theory of relativity.

These goals took into account the fact that Viking orbiters and landers were expected to continue operating over a significant portion of a cycle of seasonal change on Mars. Although it may seem that more than one photograph or water survey of a planetary site would merely be repetitious, this was not the case for Mars, for it experiences seasonal changes very much like Earth, with winter "frost" storms, dust storms with blowing sand, and other phenomena, such as erosion by wind, that bring continuous changes. In fact, major scientific gains might depend on synoptic studies of regions of interest.

Compared with the primary scientific objectives of the orbiters and landers, expectations for the entry science experiments were very briefly stated. The Primary Mission Summary document said simply, "Entry: Determine the atmospheric structure and composition." Easy to say, but to learn these things about a new planet during rapid passes through its tenuous atmosphere at two locations! What was meant to happen was an attempt to define the physical and chemical state of the Martian atmosphere and its interaction with the solar wind. In the upper atmosphere, the composition and abundance of neutral species were to be measured, along with the ion concentration and ion and electron energy distributions. In the lower atmosphere, pressure, temperature, density, and mean molecular weight were to be determined by direct pressure and temperature measurements together with data from the lander guidance systems. These data were all good scientific input, but would also serve in evaluating the design criteria and performance of the entry systems, in addition to lending valuable insight for the engineering of future missions.

The lander's scientific objectives had neither the mystique of the orbiter objectives nor the simplicity of those for the entry science experiments. They were straightforward, giving scientific priorities to the burning questions an "objective" scientist might have asked had he personally set foot on the strange new planet:

Visually characterize the landing site
Search for evidence of living organisms
Determine the atmospheric composition and its
  temporal variations
Determine the temporal variations of atmospheric
  temperature, pressure, and wind velocity
Determine the seismological characteristics of the planet
Conduct scientific investigations using the Viking Lander
  radio systems, engineering sensors, a magnet,
  and other information from the landers
Determine the elemental composition of the surface material

The order of these objectives took into account not only their scientific relevance as determined by the science teams and project officials, but also the time-critical aspects of learning as much as possible as soon as possible in case difficulties arose. The missions were planned to continue for months, but there was always that haunting possibility of premature truncation for any one of a variety of reasons. Both Vikings were launched from Pad 41 on the last days of their preferred launch opportunities. For mission 1, this allowed encounter dates on or before June 18, 1976, and for mission 2, a nominal encounter date of August 7. These dates were important to permit a good match of the retropropulsion requirements and performance limits and to stay within acceptable landing site lighting angles at the time of arrival. Since the first and most important task for the orbiters was to help the landers by providing site selection data, it was desirable that they begin these tasks immediately upon arrival at Mars.

The cruise phase of the Viking missions was defined as the period from the launch of the first spacecraft to 40 days before it was to be injected into orbit about Mars. At that time, the approach phase, which lasted through Mars orbit insertion of the second spacecraft, officially began.

Shortly after being launched from its parking orbit around Earth, each Viking spacecraft acquired the Sun and oriented its solar panels normal to it. About 80 minutes later, the biocaps that had provided hermetically sealed containers for the entry capsules were separated and allowed to float away so as to miss Mars entirely. About 2 days later, a 720° roll turn around the

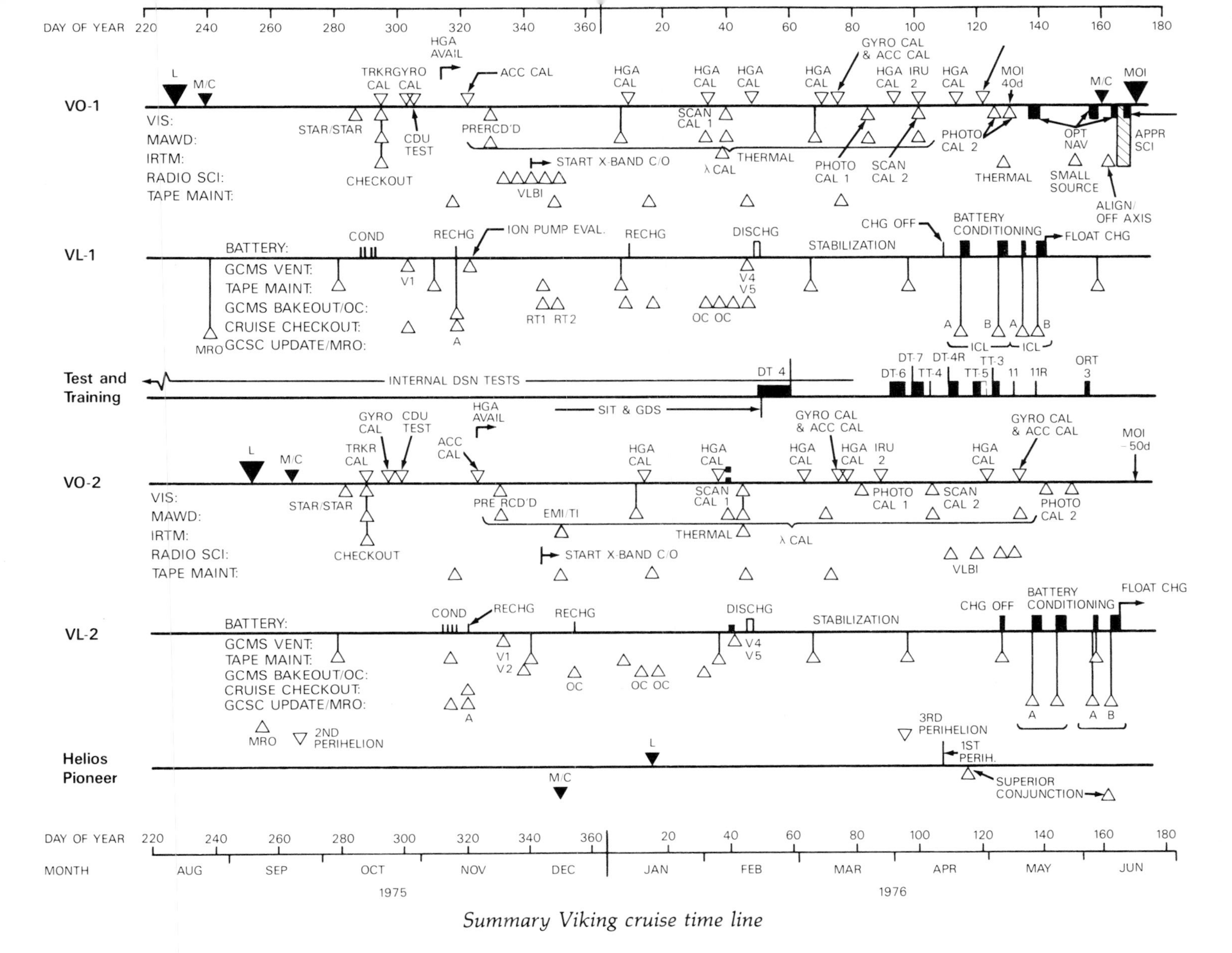

Summary Viking cruise time line

Sun axis provided for a star map and tracker calibration sequence. Several days after this, and several times throughout the misssion, gyro and accelerometer calibrations were performed to ensure that these components were working properly, as they would be needed for critical maneuvers.

An early midcourse maneuver was executed on each mission at the earliest opportunity following propellant warmup. These early maneuvers were to correct the major vehicle injection errors as soon as they were known. The first correction occurred after 7 days on mission 1 and after 10 days on mission 2. Such midcourse maneuvers were to be completed by about 30 days postlaunch, after which the orbiter propulsion system pressurants were to be shut off to prevent any possibility of leakage during cruise. Leaks of any sort would cause unwanted trajectory and/or attitude changes, much like the effects of firing small rockets.

During the early cruise periods until about 100 days after launch, communications with the spacecraft were generally maintained using the low-gain antennas, except for periods such as the midcourse maneuvers and the early scan platform and science instrument checkouts, when the high-gain modes were needed to return data. A number of engineering tasks and instrument calibrations were carried out during cruise, including venting of the gas chromatograph mass spectrometer (GCMS), battery conditioning, GCMS bakeout, and tape recorder maintenance.

Throughout the entire cruise period the Viking Flight Team on Earth was conducting personnel testing and training exercises, making sure that all players were ready for their duties when the planet was reached. A shorthand summary of the cruise activities as used by the Flight Team is included here to illustrate the manner in which the activities were integrated. There were also important interfaces with Helios and Pioneer operations; as these were occurring at the same time and sharing common DSN and SFOF facilities, close coordination between projects was necessary.

The approach phase, beginning 40 days before Mars orbit injection, signaled the start of the intense period of rapid-fire activities associated with arrival at the destination. During this period, final adjustments were to be made to the trajectories, and all science instruments and equipment that could be checked out were exercised. One midcourse maneuver had been planned for each spacecraft to finally align the flight paths and set times of closest approach; these were done about 10 days before orbit injection. The propellant supply valves that had been closed after the initial midcourse correction had to be reopened, of course, to enable the final correction to occur. When the valve was opened on the mission 1 spacecraft, a slight leak was detected which would have led to an overpressure in the propellant system after a time, and possibly to an explosion of the propellant tank. It was decided that two large-magnitude midcourse corrections, compensating in the sense that their vector sums would achieve the equivalent of the smaller correction needed, would be performed to allow the bleed-down of pressure to a safe level. To perform such complex maneuvers so near the planet without much tracking time to assess results by the time of orbit injection was a risk, but more appealing than a possible blowup. This decision clearly points out how far confidence in propulsion and guidance and control technologies had progressed since Mariner 4; the capability to perform a second midcourse correction on Mariner 4 had been added after agonizing considerations, but everyone had felt a sense of relief when the first correction satisfied the gross flyby distance criteria and the second firing was not required.

In support of the two correction manuevers, extensive observations of Mars and adjacent stars were made that provided optical data to augment the radio tracking information. Much was learned about optical guidance techniques from this activity; later missions to Jupiter and Saturn involving flyby encounters with several moons were to employ similar techniques in their final guidance input.

As the Vikings approached Mars, many observations were made of the planet to aid in calibrating instruments and checking their operation before arrival. During the period from about 5 days out to about 1 day away, a complete set of science observations was made, including color photography and global-coverage infrared measurements. The last 3 days before injection were very exciting as Viking 1 pointed its TV cameras at the Martian moon Deimos, capturing the first close-up color images of that small, rocky object against a starry background. These images also provided final optical navigation data to help in designing the orbital injection maneuver.

Because of the leaky propellant valve on the first spacecraft, the valve on Viking 2 was not opened until just before the final midcourse correction. This time things went according to plan, and only one maneuver was required. As with Viking 1, a series of approach science and optical navigation observations were made. Since Viking 2 did not reach Mars until 50 days after Viking 1, the operations teams had much additional experience to apply to the approach and landing phases for the second encounter.

Finally, after a 10-month journey, Viking 1 was injected into orbit about Mars. This began the phase designated planetary operations, signaling the beginning of the activities in orbit and on the surface of Mars. The so-called

Mission Profile Strategy for this phase was the top-level overall plan for the mission and had received a great deal of review and discussion among all interested parties, from administrators to technicians. In addition to being a menu for all activities, it was the basis for interactions among scientists, engineers, and management officials who had worked together to plan the missions and who would now be carrying them out. The Viking systems allowed for a fair amount of flexibility in operations, but there were many possible actions which were irreversible. The physics of orbits, limitations on propellants, times when viewing conditions were affected by spacecraft position, time of day, and Earth-Mars-Sun relationships were just a few of the constraints that had to be considered. The operational interrelationships of both orbiters and both landers had to be carefully regarded from the outset to avoid conflicts when multiple operations would be required.

Key factors for orbiter strategies included four considerations: propulsive maneuvers, orbit walks to relocate the spacecraft orbital parameters, identification and relative positions of reference stars, and Earth and Sun occultations that would influence attitude control and communications. Any requirement for orbiter operations had to consider these basic factors, regardless of other considerations.

For the landers, the basic strategy for operations revolved around the biology analyses and the organic analyses; these were the principal priorities for the landing mission. It almost went without saying that they should enjoy the first consideration in protocol development.

From earlier data obtained primarily by Mariner 9, a region on Mars known as the Chryse Basin had been chosen as the target area for Viking Lander 1. The selection of the initial site was based on both safety and science considerations, with safety clearly coming first. If several sites appeared to be equally safe after being surveyed, then scientific interest would become the basis of choice. A backup site was chosen on the other side of the planet in another type of geological formation, within the same latitude band, in case the primary site appeared to be questionable, but it was not used.

The safety issues that might affect the success of landing were mainly the altitude of the site, the wind conditions, and local surface hazards such as boulders. Since the atmosphere was a prime factor in providing braking during descent, the higher density at lower altitudes made them favored choices. Wind conditions were estimated by observing streaks on the surface, a very indirect indication of existing conditions, but at least providing a clue of some benefit. Areas with noticeable streaks were avoided, as were regions where surface changes had occurred since the Mariner 9 flyby. Since the best orbiter camera coverage only provided resolution of objects 100 meters in size, and since boulders greater than 22 centimeters in size could damage a lander, this hazard was dealt with by extrapolation of orbiter photos and interpretations or inferences from ground-based radar data.

While it had been hoped that the historic first landing on Mars could occur on July 4, 1976, the bicentennial of the Declaration of Independence, the 4 weeks required for reconnaissance resulted in a delay of the landing from July 4 to July 20. This was a good season to arrive, being near the beginning of summer in the northern latitudes of Mars. If organisms were present and growing, this should have been a good time to look for them.

One aspect of Mars that has always fascinated me is its similarity to Earth as a planet. It orbits the same Sun in nearly the same plane, with its rotation axis tilted 25° as compared with 23.5° for Earth, and with a rotation around its axis every 24 hours 39 minutes compared with 24 hours for Earth. This amazing coincidence (is it really?) means that Mars days are almost exactly the same as ours; but, more interesting, the tilt of its axis means that Mars undergoes seasonal changes in each hemisphere the same way they occur on Earth. Mars' orbit is farther from the Sun than Earth's; it takes about 687 Earth days for Mars to travel completely around the Sun. Thus the Martian year is almost twice as long as Earth's, which is the major difference in its general behavior as a planet.

The two approach midcourse correction maneuvers delayed the arrival of Viking 1 at Mars by about 6 hours. The site certification plan had called for the spacecraft to be over the preselected A-1 site in a synchronous orbit so that site surveys could begin immediately after injection into orbit, but the delay precluded that. An alternate plan was selected that involved an orbit with a period of 42.5 hours such that the craft would essentially overfly the A-1 site on the second orbit, allowing a retro maneuver to synchronize at that time. This alternate was executed as planned, and reconnaissance of A-1 began on the third revolution of Mars.

Although a successful orbit was a major milestone achieved, the first images of A-1 produced something of a jolt to the project team viewing them in Pasadena. When orbiter coverage of the originally chosen Chryse site was studied, many craters were evident, and it appeared that there had been extensive erosion activity and exposure of boulder fields as seen at the 100-meter resolution. The photos were more detailed than those from

Mariner 9 and showed many geologic features large enough to represent real hazards to landers. Common sense suggested that it was not a very good site.

The area to the south was known to be very rough, with deep channel beds, while images of the region to the east indicated that there had been an enormous amount of flooding in the ancient past. To the west was a vast area eroded by winds, and it almost seemed desirable to use the backup site on the opposite side of the planet.

Landing site selection (LSS) meetings had been planned all along, but they now took on a more serious character. On June 24, the first LSS meeting was held amidst exciting speculation about the preselected site; a room full of scientists and project personnel were present to see and discuss findings. Hal Masursky of the U.S. Geological Survey had been asked to lead the site selection studies, but it would finally be up to Tom Young, Mission Director, and Jim Martin, Project Manager, to decide. In addition to their own vibrations, they would rely heavily on input from Gentry Lee, responsible for mission analysis and design, and on Gerald Soffen, who, as Project Scientist, was official spokesman for the scientists.

At the June 26 LSS meeting, Gentry Lee began the discussion by announcing that the July 4 landing was in jeopardy and that a decision had to be made about whether to move the survey to alternate site A-2 or to move the search to the northwest, where scientists had hypothesized that a sediment basin might exist. Jim Martin explained that the geologic appearance of site A-1 as shown in orbiter images did not correlate well with findings from ground-based radar and that better correlation was necessary. Following considerable discussion, a vote was taken on whether to examine the northwest or to use site A-2; the result was overwhelmingly in favor of extending observations northwest toward a new site area called A-1NW.

As photomosaics of the new area were made, theories about the geology of the region seemed to be confirmed. However, the site had some rough areas, and it was not until Monday night, July 11, when the last mosaics were ready, that a site could be chosen. Jim Martin had stated that a decision had to be made the next day; he scheduled an LSS meeting to begin at 3:00 A.M. and continue until a decision was reached in the event that the issue was not resolved at the 11:00 P.M. meeting. Three sites in the A-1NW area were final candidates. After discussions of detailed studies and analyses, a unanimous vote allowed everyone to go home about midnight. Hal Masursky was able to announce to the press the next morning that the first Viking lander would be targeted for landing on the Golden Plain, Chryse Planitia, at

22.5° N, 47.5° W. Jim Martin praised the LSS process led by Masursky and the Orbiter Imaging Team, indicating that he was convinced that they had picked the safest site possible in a reasonable time.

After the landing site for Viking 1 had finally been chosen, a ground command was sent to initiate the separation and landing sequence. At this time, Mars and Earth were about 200 million miles apart, and the roundtrip time for communications amounted to some 40 minutes. The fully automated landing sequence began, first separating the lander and its aeroshell from the orbiter bus to which it had been attached for so long. After a gentle nudge by separation springs, the aeroshell-lander combination was oriented so that the deorbit rocket motor could be fired to begin the long descent from orbit to the surface. The lander then coasted for about 3 hours, gaining speed as it approached the Martian atmosphere. Meanwhile, it was sending data to the orbiter to be relayed to Earth. Just before arriving at the fringes of Martian atmosphere, some 300 kilometers above the surface, the aeroshell was reoriented for its aerodynamic entry. Its ablatable heat shield protected lander systems from the intense heat, decelerating the lander to a speed of about 250 meters per second so that a parachute could be deployed from a can by a small mortar. This device, 50 feet in diameter, had been packed into a small container well before the launch, much longer than the 90- to 120-day maximum normally specified before repacking is required for emergency parachutes. The parachute essentially pulled the lander away from the aeroshell, allowing it to drop to the surface, as it slowed the lander to about 60 meters per second some 1.5 kilometers above the surface.

At that height, a marking radar called for the firing of three retro rockets mounted directly on the lander spacecraft. These engines burned for about 40 seconds, being throttled by commands from the computer, based on sensor information from the radar system. The last 30 meters of altitude were covered with the spacecraft descending vertically in a gentle fall at about 2 meters per second. A switch on a landing footpad signaled shutoff for the rockets, and Viking landed.

Landing sites had to be elliptical in shape, about 100 by 300 meters in size, to allow for uncertainties in control over touchdown. Lander 1 touched down within 20 meters of the center of the chosen ellipse, so the "guesses" of the engineers must have been better than expected.

During the design and development period for the landing rockets, there was concern about the effects of jet blasts on surface materials. Simulation firings of motors in dust led to studies of multiple nozzle concepts and the

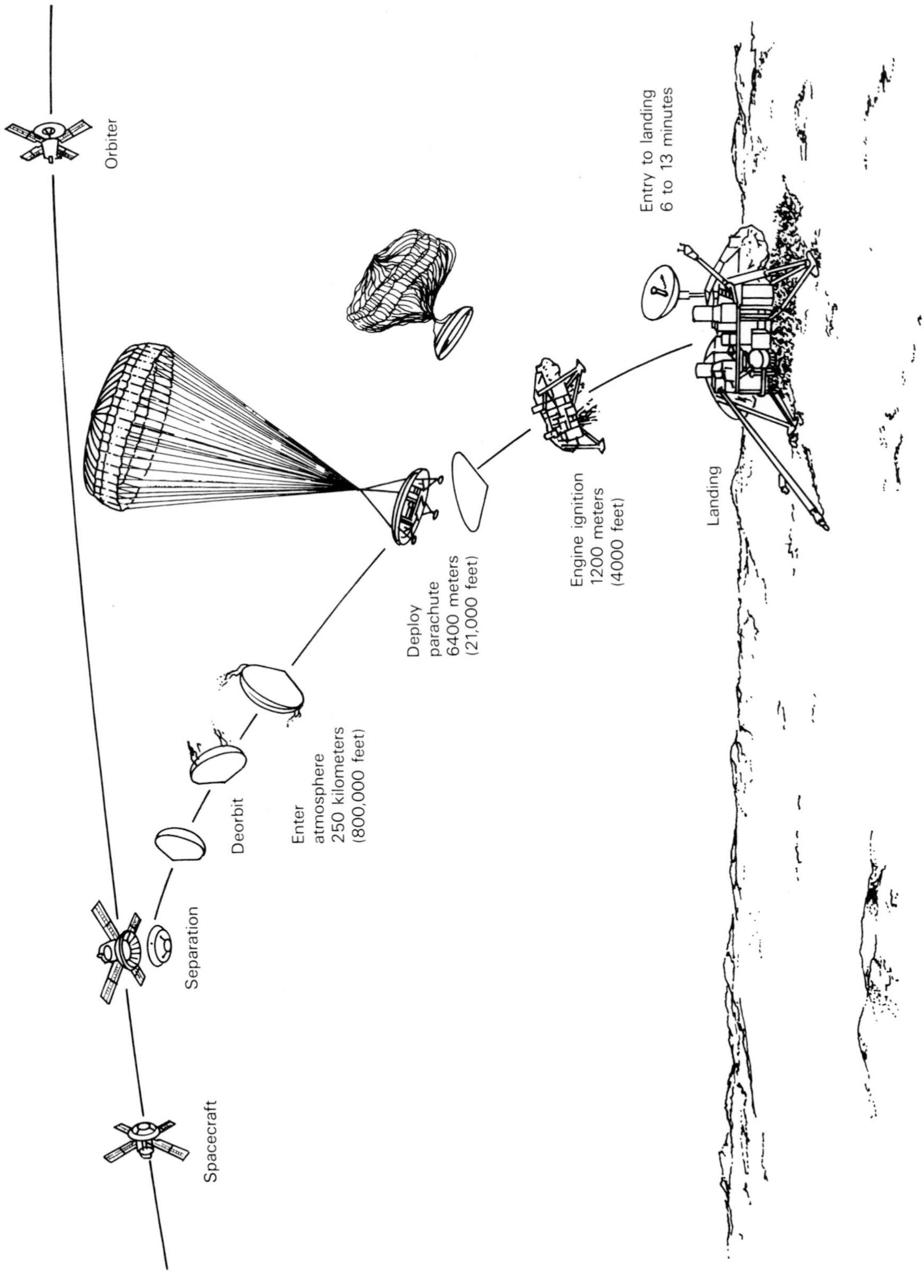

Spacecraft

Separation

Deorbit

Enter
atmosphere
250 kilometers
(800,000 feet)

Deploy
parachute
6400 meters
(21,000 feet)

Engine ignition
1200 meters
(4000 feet)

Landing

Entry to landing
6 to 13 minutes

Orbiter

Viking entry system

development of an 18-nozzle rocket that showed minimal effects from impingement on the surface. Many tests were run to determine temperature, chemical, and mechanical problems that might be induced by the motors. To make a long story short, the effects were well analyzed, the multiple-nozzle rockets were successfully developed, and no known problems occurred as a result of using rockets to achieve the soft landing.

During the descent through the upper atmosphere, entry science instruments in the aeroshell made measurements of the ions and electrons in the upper atmosphere and the neutral species in the lower atmosphere. Pressure, temperature, and acceleration measurements were also made during descent. All this happened in about 10 minutes, but it was another 20 minutes before those of us patiently waiting in the SFOF knew that it had been accomplished successfully.

During the entry and landing period, I was stationed in the "glass cage" shared by mission directors Tom Young and Bob Crabtree. The SFOF was crowded with the Viking project members who *had* to be there; with visitors like me from Headquarters, Langley, and corporations that had helped build Viking, the place was packed. Several "bullpen" areas housed engineering specialists with their tightly spaced desks, videomonitors, and telecom consoles. Surrounding these were the glass offices occupied by management officials and their special display and communication systems. The glass provided some shield from noise, but allowed almost all of the operating team to view the comings and goings of colleagues. It was a scene of high technology communications activities, but I was amused to see it occasionally augmented by a frantic wave, by pointing, or by some other primitive, human hand signal used as an expedient.

I could think of no place I would rather be during the final minutes of the first landing on Mars. The two mission directors had as much real-time information available to them as anyone on the team, and I had developed the highest respect for their competence over the many years we had been working together. Bob Crabtree had been involved in the operations activities at JPL and at the Cape from the very early days of Mariners 1 and 2, quietly advancing to more responsible positions until he was leading Viking orbiter operations. I had watched Tom Young develop from a mission integration engineer during the successful Lunar Orbiter program to his present very critical position as Mission Director for Viking. We were sure to have the facts as soon as they were known, and I was proud to be with two of the key "Vikings" in this crucial period.

Words seem inadequate to describe the last 10 minutes of the landing sequence. I may have been in a state of suspended animation, although I thought at the time I was just being cool. After all, I had been through this process seven times before during Surveyor landings on the Moon. I don't think Surveyor even entered my mind as the key events of chute deployment, aeroshell jettison, engine start, and velocity-altitude callouts occurred, followed finally by the indication that the telemetry bit rate had switched from 4000 to 16 000 bits per second. This signal, 10 seconds after touchdown, was the mark of survival; had the landing been destructive, the signal would not have been given. It seemed almost too good to be true, but then we were quickly caught up in the handshakes and backslapping of mutual congratulations. This made the time pass quickly, and before long the festivities were interrupted by the appearance on the monitors of the first linescan image of Mars' rocky surface and a Viking footpad, clearly resting solidly on the target planet more than 200 million miles away.

On the heels of the thrills that went with the successful landing and remarkable pictures came one other small personal experience I will not forget. It was a result of my wandering around visiting with friends everywhere in the SFOF during the long period between lander separation and entry into Mars' atmosphere. I entered a glass cage where Israel Taback of Langley, John Goodlette of Martin, and other systems experts were waiting for the next events. Taback had been a respected friend since Lunar Orbiter days and had functioned as chief engineer throughout the Viking effort. He met me with a broad grin, asking if I would like to get in the pool. I immediately recognized this as a sucker setup, with me as the sucker, but I naturally responded with, "What pool?"

"The blackout pool," he said, meaning a pool for guesses as to how long the radio blackout would last as the entry capsule passed through the atmosphere. Just as communications blackouts always occurred when spacecraft were returning to Earth, this same phenomena was expected to a lesser extent for Mars. Naturally these men had been thinking about this effect as part of their jobs and, for all I knew, had the benefit of some astute calculations to support their guesses. Nevertheless, since the amount of the "donation" was only a dollar or two, to enter into the spirit of things, I joined the pool.

When faced with the challenge of picking a number, I suddenly had a hunch that the very tenuous atmosphere and the conservatism of the communications engineers, who always had a surprising amount of margin in their predictions, might just combine to result in no perceivable blackout at all. The rest you can guess; my "zero" blackout time won, and Taback sheepishly came to me with a handful of bills and the secret ballots that had been cast by "us ionospheric physics experts" who were confirming "prior" knowledge of conditions at Mars.

Almost like the chrysalis transition from a caterpillar to a butterfly, Viking became a different object upon landing. What had been, moments before, a flying machine of the most sophisticated sort, was now a scientific laboratory, immobile and wedded to the soil. Changed, too, were some of its masters, for there were no longer any tasks involving rocket or attitude control functions to perform, and those who had played such a vital role in getting to the surface of Mars were no longer needed. The Viking Lander 1 was now a laboratory dedicated to the conduct of premier scientific investigations of historic proportions. New masters came forward to command it.

There were still many engineering functions to be performed and monitored for the Viking laboratory, just as there are for laboratories on Earth, which demand engineering support to operate as effective facilities. Such functions were not unimportant; they simply assumed a different priority when the scientific investigations began.

The lander's presence on Mars also brought about a subtle change in our thoughts about time. The lander was now a creature of a new world where the days were 24 hours and 39 minutes long. Not much different from what it was used to, but enough to accumulate over a period of time and make the sunrises and noontimes change. If it was to operate as an entity studying the environs of this strange place, it would have to operate on local time, the same way you or I would adjust in a foreign land. And so, too, would its masters on Earth have to adjust.

To deal with operational time based on a Martian day, the term SOL was invented. For a long time I supposed it to be another acronym that I needn't bother to learn, but I haven't found anyone who knows how it came to be. As a substitute definition for a "Mars day," these SOLs became the units of time for planning all operations on Mars, even the work shifts of the people involved in lander operations. However, their days were so unroutine that there never was such a thing as a 9 to 5 shift geared to SOL. The entire operation was, at best, on flextime; realistically, it was probably more like continuous overtime.

While everyone else was gasping and exclaiming over the images coming in on their screens, the entry science team was pouring over the data they had received concerning the atmosphere. A major question had arisen over argon content, primarily because of a Russian estimate that argon made up 35 percent of the atmosphere. If the Russian estimate had been correct, the GCMS instrument might have had difficulty. After a few hours of study, the results clearly showed that the argon concentration was only about 2 percent, a much more reasonable number and no cause for alarm.

The folded meteorology boom, carrying sensors much like those seen on a small Earth weather station, extended soon after launch. Right after the first look around with the cameras, Viking did what any new arrival would have done and observed the status of the weather.

The meteorology team gave their first Mars weather report early the next morning, telling us that there were "light winds from the east in the late afternoon, changing to light winds from the southwest after midnight. Maximum wind was 15 mph. Temperature ranged from −122° F just after dawn to −22° F [but this was not the maximum]. Pressure steady at 7.70 millibars." This was the first of a daily (SOLy?) series of reports from the Viking lab on the changing weather conditions on the surface of Mars. Quite a while later, a morning report told of a winter storm that was verified by camera images showing what appeared to be light snow covering the ground. The meteorology instruments worked well on both Viking landers, and we soon had enough seasonal data on Mars weather to joke about a Martian Farmer's Almanac.

The next instrument to be activated was a seismometer. While not a primary instrument in the search for life, it was expected to provide basic information about the origin and evolution of Mars. Efforts to uncage the instrument were disappointing, and troubleshooting did not succeed in getting it to work. Fortunately, its counterpart on Viking 2 performed flawlessly, so that data about Mars quakes were obtained. Mars appears fairly inactive seismically, and most of the disturbances measured were believed to be related to the effects of wind.

As if to prove its human qualities by showing that "nobody's perfect," a command to the surface sampler control assembly (SSCA) caused the collector head to retract too far, crunching a restraint or latching pin and inhibiting its release. This caused a stir in Mission Control, for the surface sampler was vital to the biology and chemistry experiments. The SSCA was the "arm and hand" that had to reach out and collect samples, pick them up, and load the hoppers of the "chemistry lab," whose exotic and unique instruments would have been useless without them. This brings to mind the dependence scientists must often place on technicians who serve loyally in laboratories on Earth, as well as the fact that they are often taken for granted—until they do not respond as expected or make mistakes. A further object lesson of the Viking surface sampler experience came from the troubleshooting activities that were begun immediately, for they showed the problem to have been caused by an incorrect command, not an improper response.

A diligent effort on the part of a team led by Len Clark of Langley quickly determined where the trouble was and how to remedy it. This involved reworking the commands and checking them carefully on the prototype hardware used in simulations. When everything was ready, including new commands for pointing the cameras to observe pin release and the location of the sampler after it was dropped to the surface, the new instructions were sent and executed perfectly. All these unplanned activities meant that the SSCA was unable to perform until SOL 5 instead of SOL 2 as scheduled, but everyone was so relieved to have things right again that Clark was congratulated for his heroic effort.

The development of the surface sampler arm and hand involved a combination of electromechanical technologies and the fundamental physics of its human counterparts. It had to be capable of being stowed out of the way until after the landing, able to extend in the desired direction, reach surface features over a nearby circular arc, manipulate the surface, pick up samples, and place them in receivers that allowed the samples to be processed. We humans take our arms and hands for granted, but those who have contemplated the design of their replacements are well aware of the magnificent sophistication of the combination of sensors and the mechanical and control systems involved. Compared with the dynamic, adaptable qualities of a human arm, the SSCA was extremely simple; nonetheless, it was effective in the Martian environment.

The SSCA could extend 13 feet from its mount, reaching the ground from 3 to 10 feet from the spacecraft. Its radius of operation was about 120°, giving it a surface area coverage of about 95 square feet. Normally it would be programmed to the desired azimuth, extended the desired amount, and lowered to the surface. It would then extend into the soil about 16 centimeters with its jaw open, acquire a sample, retract the collector head with the jaw closed, and then elevate and deliver the sample to the desired receiver. Of course it could be used in other ways: as a trenching tool, as a rock hammer, or to move a rock on the surface. In addition to its scoop and backhoe, it carried magnets and a thermocouple to determine magnetic properties and surface temperatures. Its motor load could be recorded to give an idea of the cohesion of the soil or the forces required to scrape a sample. A vibrator was installed so that loose soil samples could be shaken loose from the collector head if necessary. Though limited to "feeling" Mars in a very narrow area, the surface sampler did provide this surrogate capability for the scientific explorers who commanded its actions on Mars, and it served its technician role faithfully as long as the commands given to it were appropriate to its basic capabilities.

In some ways, the design and operation of the surface sampler presents an eerie parallel to applications of prosthetic devices. Its substitution for a human arm and its special requirements for control and manipulation required skill and learning. A mission status bulletin in October 1976 described the tricky process of pushing rocks 200 million miles away. This experience with Viking Lander 2 resulted from the desire to learn more about the nature of the rocks and Martian surface materials by moving some of the rocks near the lander. These had all been observed in their undisturbed state for a period of time; now it was reasonable to turn one over and to really see what it was like. Although it had taken longer to get this far in the exploration process because of the remote operations, the action was not different from that an explorer might have taken had he been able to use his own hands and tools. The account in the bulletin read as follows:

> Well, we tried one and it didn't . . . and then we tried another and it did—but not far enough. So we did it again. That's the way the first rock-pushing event proceeded. The first attempt was with rock 1-ICL rock. A nudge was successfully conducted October 4, but there was no budge with the nudge! ICL rock was, one might say, "unmoved" by the Lander's attempt to dislodge it from its place of rest—a place it has probably occupied undisturbed for perhaps millions of years.
>
> The conclusion is that the rock is like a floating iceberg—most of it buried and simply too large to move. Not to be foiled by this kind of development on a planet that seems to take pleasure in confounding our spacecraft (we're getting wise to its tricks), a second sequence was already in the system to transfer the soil sampler's affections to rock 3—and that push was successfully conducted October 8 on the same schedule planned for rock 1. But the game wasn't over yet.
>
> Have you ever tried to push a good sized rock out of the way in your garden with—say—the handle end of a hoe or some other kind of long implement with a small contact area at the pushing end? Unless the contact point is very precisely centered on the mass of the rock, when the rock is pushed it might do any number of strange things. These physical laws still apply on Mars, and the boom-extended surface sampler is not immune to their effects. During the first push, it appears that the rock rotated more than it moved in a straight line, and there was agreement that it

had not been moved far enough. This was a "trick" we weren't prepared for, and it is a tribute to the designers of the surface sampler command sequences that they were able to get a new sequence designed and implemented in very short order during the weekend so that the new push could be carried out October 11. The pictures illustrating their success were received that evening. Rock 3 not only moved adequately, but lifted out of its semi-buried position to reveal itself to be more than twice its original visible size.

Returning to the problem with the surface sampler pin that was solved on SOL 5, it was SOL 8 before everything was ready and a trench was dug by the sampler. A sample of Martian soil was transferred to the chemistry lab, technically described as the biology, gas chromatograph, and X-ray sample processing and distribution assembly. The most interested "Vikings" gathered at 6:30 A.M. to watch, as images of the trench dug by the sampler were returned slowly on the direct video link from the lander. The surface looked appropriately disturbed, and presumably the samples had been delivered. Engineering data showed that the biology instrument had obtained enough of a sample, but something had apparently gone wrong with the processing for the GCMS. Evidence initially pointed to a low level or insufficient sample size, but no one knew why that could have occurred. Needless to say, this caused a great deal of worry, and theories were developed about what had happened and how to recover. Three possibilities were considered: (1) no soil had been delivered to the distribution assembly, (2) the level-full indicator had not operated, or (3) some unusual quality of the soil had kept it from flowing through the distributor.

After much discussion, it was decided that a reasonable course of action was to collect another sample on SOL 14, disable the level-full no-go signal, and attempt to perform experiments on the available sample, even if the quantity was smaller than desired. Of course, the TV was used to observe the sample site, the processing and distribution assembly, and the dumped sample remaining in the collector head that did not pass through the sieve. Following this day of work, the spacecraft would be commanded to analyze the sample on SOL 15 and again on SOL 23. The limited amount of "equipment and expendables" contained in the Viking chemistry lab plus the extreme care necessary to ensure proper commands and interpretation of data made it prudent to proceed slowly.

Meanwhile, rumors flew as the scientists met on SOL 11, for the biology team had received its first data and was to report on the results. Vance Oyama from NASA Ames described his findings from the gas exchange experiment, in which water vapor was released into the soil and evolved gases were analyzed to determine whether organisms would exhale exhaust products. His results indicated that the soil was active in some way, but he was cautious about the reason there seemed to be such a large amount of oxygen produced. This was the kind of happening to be expected if plants were growing, but . . .

Data from the labeled release experiment described by Pat Straat were even more exciting. In this experiment a radioactively marked nutrient that might be "devoured" by living organisms was introduced into the soil. By monitoring the amount of radioactive carbon dioxide produced, an idea could be obtained about the quantity of such organisms. Results from this experiment showed an incredible amount of activity. It was possible to interpret these results as indicators of both plant and animal forms of life, in significant abundance, but soon after these presentations another possible interpretation was offered by John Oro, a member of the Molecular Analysis Team. He suggested that it might be possible to obtain such results if Martian soil contained peroxides that were decomposing to release the oxygen detected in the gas exchange experiment. Such chemical constituents could also break down the formate and other organic ingredients of the labeled release nutrient, causing the large signal suggesting carbon dioxide. This somewhat unusual theory was reluctantly accepted as a possibility pending future analyses, but it dampened hopes enough to keep reports of life on Mars from being blown out of proportion in the newspapers.

There was hope that the matter might be settled by detailed analysis of data from the gas chromatograph mass spectrometer, which would be able to determine the presence or absence of organic compounds or carbon molecules. When, on SOL 14, the sampler again performed its task, the level-full indicator did show positive results; a sample had been delivered to the GCMS soil processor. Klaus Beimann presented bad news for the biologists, reporting no evidence of organic substances in the soil. This was the beginning of the negative results concerning evidence for life, as repeated experiments and analyses seemed to confirm the theory that chemical combinations—not expected based on Earth conditions—were responsible for the active biology indications.

Many of us were very disappointed. It had been a little like waiting for Christmas as a kid, only to find on Christmas morning that Santa didn't come through. All those years of talking about the possibility of life and planning to find out about it had built up the tremendous hope that exciting results might be obtained. But after thinking about the negative results for a long time now, I have replaced those disappointments with other thoughts perhaps more exciting. Maybe the fact that there appears to be no life on Mars is better than finding that there were only some low forms of algae or distorted microbes of little significance. Mars may be a pristine site for future development through "planetary engineering"; perhaps we can make it an habitable and otherwise useful place for expansion. After all, the future residents of Mars might not have to worry about ants at their picnics or contend with viruses that have a long history of development.

Such biological speculations are far beyond my qualifications, but I will be ready to consider how we might create an atmosphere and satisfy other environmental requirements to enable a manned expedition to succeed. The fact that the planet represents a lot of real estate—not much more bleak than the deserts of Arizona—makes it a "place" to explore and examine for its intrinsic value. It has mountains, canyons, large volcanoes, and other features that are magnificently awesome. What little we have deduced about its history indicates that it has seen dynamic periods of flooding and erosion that might have been accompanied by spectacular developments—unimaginable, almost, until one reads about the continental drifts, the ice ages, the dinosaurs, and other stranger-than-fiction occurrences on our planet. What might we find if we could roam the Martian surface and dig around? After all, its land surface area of 55 million square miles is equal to about 40 percent of the land area of Earth, and we have only been able to scratch 180 square feet with our samplers.

Spinners Last Forever

Not all interplanetary travelers are like the Vikings, sophisticated caravels sailing across vast space, stabilized in three axes. Some are from a different family: spin-stabilized spacecraft that trade the complexity of three-axis attitude control systems for the simplicity of constant rotation.

Often thought of as no more than a toy for 10-year-olds, the gyroscope is actually a subtle and versatile inner ear for machines, providing attitude reference and control where nothing else can. Gyroscopic principles are used in all manner of devices, from bicycles to nuclear warheads. In a ship's wheelhouse a gyro supplements the traditional magnetic compass, using inputs from this ancient instrument given to flighty and deviate behavior, and making it useful for precise navigation. Aircraft instrumentation is rich in gyros, notably in the automatic pilot that relieves the human pilot from the constant attention needed to fly a steady course in a turbulent medium. In submerged submarines gyros are part of a marvelous machine that senses every change in heading and every variation in speed and current, integrating the multiple variables with such precision that the skipper, although functionally blinded, can know his exact position after weeks without a conventional real-world fix. In ICBM guidance systems, gyros endure a high-G launch, arc a thousand miles upward into space, survive incandescent reentry, and guide their warheads wickedly to target.

These feats, which range from the everyday to the apocalyptic, are performed by sensitive, mulishly independent mechanisms that use concepts defined by Isaac Newton to do things no mortal could manage unaided. In the delicate tasks of interplanetary navigation, gyros have earned two quite different classes of duty.

For spacecraft that are stabilized in three axes by sighting on distant objects, it is periodically necessary to give up this cruise orientation and slew to a different attitude before firing trajectory-correcting rockets. Gyros in an attitude reference package allow this to be done precisely, maintaining reference coordinates all the while. After the velocity corrections have been made, the spacecraft may be reoriented to its original cruise attitude. For all these tasks, gyros serve nicely, keeping the control computer aware of which way is "up" in a universe without up.

The other application of gyro principles to spacecraft function is of a different order. If the entire craft is made to spin, it becomes in effect the rotor of a large gyro and is thereby stabilized in inertial space along its axis of rotation. Although it has drawbacks, this is a long-lived, low-energy way to keep a spacecraft oriented during its travels.

The principle by which a gyro works seems uncomplicated, yet its reactions to external forces are mysterious. Spin a wheel and observe that the axis on which it turns has gained an odd property. It resists deflection and "wants" to hold position against side loading. But if you overcome this resistance and compel it to point in a different direction, note that, unexpectedly, it precesses and "wants" to move in a plane 90° to the deflecting force. (This is what gives so curiously animate a feeling to a handheld toy gyro, like a little animal trying to escape.) Enclose the spinning wheel's axis in a polar hoop, and then enclose that hoop in an equatorial one, and you have the heart of a neat device capable of keeping its orientation in inertial space.

Of course, the realities of applying simple physical principles to machines can be difficult, and the gyro application invites complication. Much skilled instrument engineering has gone into gyros to make them practical, rugged, and reliable. Further effort has been devoted to attacking a constitutional sensitivity to external forces: in time the heading established by a gyro drifts into error. No matter how carefully the instrument is made, it remains susceptible to the accumulated effects of tiny forces caused by bearing friction, temperature fluctuations, or even the presence or absence of small magnetic fields. Over time, these influences add up to error. In recent years the limitations of mechanical gyros—never so great as to impair their usefulness over moderate intervals—has been moderated by an exciting development, the laser ring gyro. In effect these gyros are made by replacing the rotating mechanical parts with rings of laser light, rotating without friction. Each laser gyro consists of two rings of light traveling in opposite directions; motion causes the frequency of one beam to be upshifted and the other downshifted. The sensitivities are such that changes in rotation at the rate of 10° an hour cause a detectable frequency shift. These devices are finding application as mechanical gyro replacements, and new orders of accuracy and stability can be expected when they fly on interplanetary errands.

From the earliest days of rocketry, spin stabilization has been employed during the rocket burn. Just as the feathers on the shaft of an arrow or the rifling in the bore of a gun provide spin to stabilize a projectile, spacecraft are often mounted on final-stage solid propellant rockets that are spun to give a fixed thrust direction during burn. After rocket burnout, the spacecraft may remain attached or may be separated, in either case continuing to spin about the same axis. If the spinning is undesirable, or if the rotation rate must be changed, despinning is achieved by a simple technique of unwinding and then releasing small yo-yo weights.

JPL engineers still recall one early Explorer that successfully progressed through a multistage launch, all going well (which was remarkable for those days), until the spacecraft and its final stage achieved the desired trajectory. Then, thanks to a certain prelaunch oversight concerning the moments of inertia, the spin axis changed from the longitudinal axis of the launch to 90° from this axis, where the small vehicle was actually more stable. The laws of physics were still perfectly obeyed, but this embarrassing bird preferred to spin sideways. It was an instructive failure, about the only salvation of the experience. A related event occurred several years later when a more expensive advanced technology satellite was tipped on separation and spun in a direction opposite to the intended one, making it impossible for its yo-yo weight system to unwind and stop its spin. In this case, the sure-fire aspects of spin stabilization will forever haunt engineers.

Spinning an interplanetary vehicle to provide orientation in space has several implications that deserve discussion. One arises from the need to manage scientific observations in some uniform fashion. A spinner with sensors looking outward radially will sweep the sky in a systematic and predictable manner. As the spacecraft orbits its parent body—the Sun in the case of most interplanetary vehicles—these swaths of coverage can be predicted and counted on to view the interplanetary medium on a regular basis. For measurements of magnetic fields, radiation background levels, and similar spatial information, this controlled scanning mode has clear merit. Of course, for a planetary flyby, where the desired look angle is much less than 360°, a spin mode offers few advantages, even though, as will be noted later, it can be employed. But for interplanetary observations, the scanning qualities of a spin-stabilized spacecraft are useful.

A second factor affected by the stabilization of an interplanetary vehicle is the generation of solar power. With three-axis stabilization it is possible to position arrays of solar cells perpendicular to the Sun, the most efficient angle. With a spinner, the designer must settle for somewhat less, even though some arrangements are entirely practicable. If the spin axis is normal to the plane of the ecliptic (the plane occupied by the Sun and planets), then a cylindrical spacecraft having a band of solar cells that encircles the spin axis will be oriented so that the Sun serially illuminates all cells, creating a continuous ripple of power. Of course, more cells must be carried for such a cylindrical array than for a simple planar array, since the entire band of cells is never illuminated at the same time.

The third aspect of spinners to be considered involves communications to and from Earth. The earliest spinning spacecraft used low-gain, omnidirectional antennas, handy if some mischance tipped or canted the spacecraft into an unplanned attitude, but less than desirable for a large volume of error-free communication. As the two-way data link to Earth was of critical importance, higher-gain antennas that produced fan-beam, focused patterns were developed; if aligned so that the pancake-shaped beam intercepted Earth, they were not affected by the spin.

Aiming the antennas of Earth-orbiting satellites toward Earth presented small problems, but the geometry grew trickier when spacecraft were dispatched to the farther reaches of the solar system. The problem arose in the design of the second block of Pioneers, designated 6 though 9, sent into solar orbit to examine the interplanetary medium. Unlike the first block of Pioneers, which, except for Pioneer 5, were early lunar probes plagued by erratic launches and unreliability, this second block, launched from 1965 to 1968, were uncommonly successful spacecraft, reliable and richly rewarding in scientific return. The antenna-pointing problem could have been severe, as these birds were put into solar orbits not unlike the Earth's, but trailing or leading the home planet by large fractions of its annual path. They used a Franklin array antenna that transmitted and received signals in a fan-shaped pattern oriented to include both the Sun and Earth in its coverage of the ecliptic.

It may be well to examine how constantly spinning spacecraft can be adapted to the imaging of objects in space. Several ingenious methods have been used: one employed in a final block of Pioneers, the highly sophisticated Pioneers 10 and 11, made use of an instrument known as an imaging photopolarimeter. Looking radially outward from the spin axis as the spacecraft flew past a planet, it collected a narrow swath of image information as spacecraft rotation caused it to scan the target. On successive rotations an adjoining swath was viewed by slightly adjusting the field of view, and so on until the entire planet had been imaged. The swaths of light data would be transmitted serially to Earth and reassembled into a single image. Putting this simple principle into practice involved sophistication depending on the geometry of the flyby, the prevailing angle of illumination, and the areas of particular scientific interest. However, as the beautiful Pioneer pictures of Jupiter testify, it proved entirely workable.

A different approach to compensating for the inconvenience of spinning instruments was used on the Orbiting Solar Observatory satellites in the 1960s. A separate, free-turning portion of the spacecraft was made to spin while an instrument-carrying portion was oriented relative to the Sun (the object being viewed). The gyroscopic forces on the spinning portion thus maintained orientation, and, in the weightlessness of orbit, the forces on the connecting bearing were minimal, so that friction was not a significant factor in maintaining the spin rate of the rotating section.

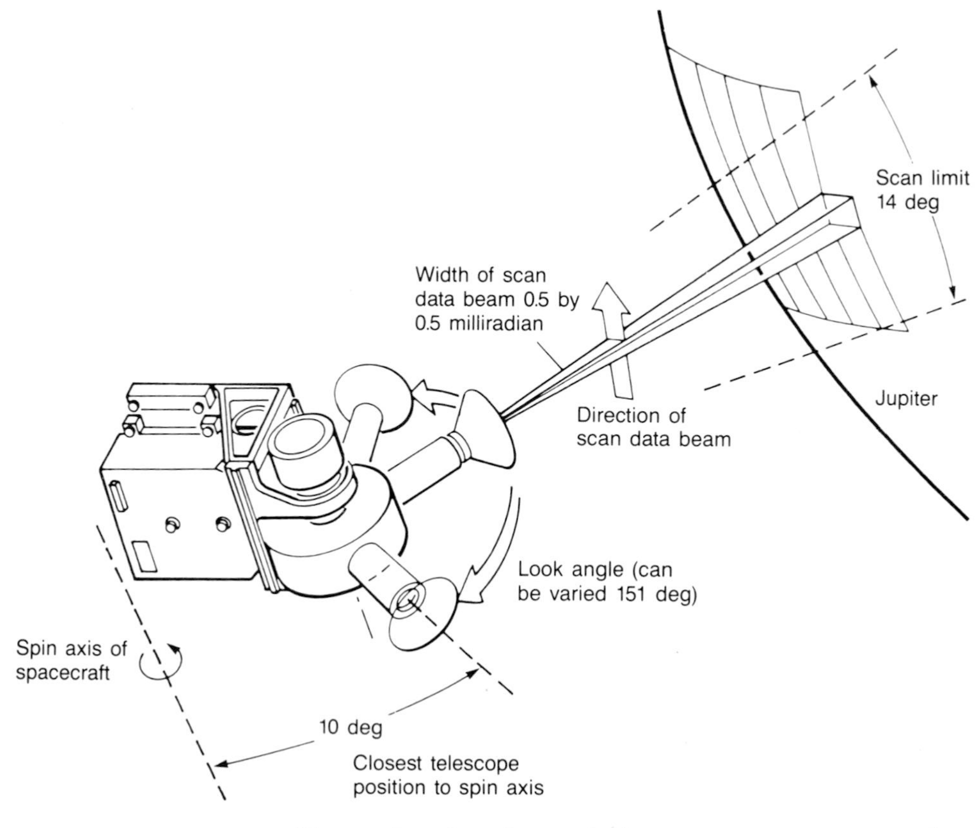

Pioneer imaging photopolarimeter

The engineering problem of carrying multiple electrical connections across the spinning interface was solved by using slip rings made with exceptional quality and precision. However Rube Goldbergian they may have seemed, the Orbiting Solar Observatories worked nicely in orbit, which was what counted. The concept of the two-part spin-stabilized spacecraft is destined to fly again when the Galileo spacecraft, scheduled to study Jupiter in the late 1980s, will be spin stabilized, with a nonrotating instrument platform.

Although they never won much public attention and respect, the early Pioneers were interesting spacecraft. The first one, launched in August 1958, suffered the misfortune of a flawed first stage that failed; it became known as Pioneer Zero. It nevertheless lingers in the memory of Charles P. "Chuck" Sonett, then a scientist nursemaiding a magnetometer aboard the craft. Just before launch, he climbed up the gantry for a last look at his instrument. Horrified to find that a vital shield had come loose, he hurried down, borrowed a soldering iron, and was starting back up again when he was stopped by an imperious safety officer. "You can't plug that thing in," he was ordered. "We've got live rockets stacked here." Expostulation was useless. A technician found an electrical outlet away from the rocket, heated the soldering iron, unplugged it, raced up the gantry, made a few dabs at the loosened shield until the iron cooled, scurried back down to reheat the iron, and repeated the process until the shield was secure. The valiant effort was futile, of course; the rocket failure launched the spacecraft to disaster.

Three months later Pioneer 1 was launched. It failed to reach the Moon, its nominal destination, but it did return 43 hours of data about the then mysterious interplanetary medium. It is not easy to recapture the extent of our ignorance a quarter-century ago; *everything* we learned was new. The first four Pioneers had been planned as lunar reconnaissance spacecraft, at which they failed; Pioneer 4 achieved the highest orbit, approaching within 37 300 miles of the Moon and sending back significant quantities of interplanetary data. It was this series of spacecraft that greatly advanced the definition of the Van Allen and other radiation belts in the vicinity of Earth, following their initial discovery by Explorer 1.

Pioneer 5 had originally been planned for a possible flyby of Venus but was not ready in time for launch at the planetary opportunity in late 1959. It did achieve a solar orbit and became the first spacecraft to send data back over a distance of 22.5 million miles, the longest radio transmission distance achieved at the time. The information that it transmitted from March through June 1960 fascinated interplanetary scientists by revealing temporal and spatial variations of particles and fields in the region between Earth and the orbit of Venus.

This series of spinners—Pioneer 0 through 5—was begun prior to the creation of NASA and was the continuation of a program started in the earliest days of U.S. space development. With NASA attention turned toward Rangers, Surveyors, Mercurys, and a full complement of physics and astronomy satellites, the appetites of a small but increasingly interested cadre of interplanetary scientists were whetted just when the outlook for future interplanetary launches disappeared.

Having been heavily involved in the early Explorers and Pioneers at Space Technology Laboratories in California, Chuck Sonett was a leader in the interplanetary field. He came to work at NASA Headquarters in November 1960, bringing not only a strong scientific background and understanding about fields and particles in interplanetary space, but also a significant amount of engineering experience in the design of instruments and interplanetary spacecraft. His early attempts to satisfy the increased interests of interplanetary scientists with instruments riding on Ranger and Mariner spacecraft resulted in frustration, because of the priority conflicts in the selection of scientific objectives. Experiments aimed at gathering new information about the Moon or a planet at arrival always seemed to receive priority over those examining the interplanetary environment. This resulted in compromises that prevented orderly planning and acquisition of interplanetary facts.

The success of the early Pioneers, although modest, was enough to convince Sonett that special *inter*planetary spacecraft were a much-needed element in a total program, rewarding not just for their return of interplanetary data but also to support the engineering modeling and design of spacecraft that were to journey through space to other planets. Many questions remained about the radiation environment and its effects, especially transient energetic events like solar flares, and about such ill-defined factors as micrometeorites and magnetic fields. At the time, data did not exist to properly model the solar constant at distances related to the nearer planets.

This special interest in interplanetary study eventually became a major factor involving the Ames Research Center, a NASA laboratory that previously had played a large role in developing reentry aerodynamic concepts, but which had not become a participant in space project activities.

When NASA was created and former NACA laboratories became heavily involved in space projects, there was a great deal of change and, some thought, erosion in existing research activities. This was a concern to NASA's Deputy Administrator Hugh Dryden and to Ira Abbott, who headed the office responsible for advanced research. As a result, Headquarters established guidelines that encouraged research and development work at Ames, Langley, and Lewis, with minimum dilution from space project activities. Langley had already undergone a significant transformation to manned space activities, with the assignment of a Space Task Group, resulting eventually in the transfer of some 250 researchers to Houston. Several key personnel from the Lewis Research Center had come to Washington to help staff the space flight organization under Abe Silverstein. Only Ames had failed to undertake any major space project after 2½ years as part of NASA.

By this time, the Goddard Space Flight Center had been assigned a principal role in Earth satellite projects for physics, astronomy, and applications areas, JPL was up to its ears in lunar and planetary programs, and the options for new efforts were limited. Furthermore, senior management officials at Ames and at Headquarters did not seem impelled to strain against the "avoid diluting research" guidelines.

This view was not shared by a small group of engineers and scientists at Ames. They were specifically interested in the Sun and its effects on Earth, and they conceived a solar probe that would travel inward toward the Sun and be ideal for making interplanetary measurements. The technical requirements for a spacecraft that was to operate in an extremely hot environment could be studied with facilities existing at Ames and appeared to be a good match for their scientific talents. Like the other NACA laboratories, Ames had an unusual array of talented people who had been working in high-technology areas on the fringes of space for years and were ready to contribute more than research support to the rest of NASA. Names like Harvey Allen, Alfred Eggers, and Al Seiff were synonymous with high-temperature, high-speed flight. Harvey Chapman had made planetary entry trajectories and other analytical determinations easier, and many engineers at Ames understood the physics and chemistry of aerodynamic heating better than most.

Charles Hall came to Headquarters to make a presentation in December 1961; Ames engineers had done their homework toward defining a good solar mission, and it was evident that the group very much wanted to become involved in the project management of a space mission. At the same time, plans were underway to define experiments in support of an International Quiet Sun Year, and there was interest in a meaningful mission.

At Headquarters we were interested but wary. While the project could fulfill a basic scientific need, and the Ames engineers had distinguished themselves in research activities, none of them had obviously relevant project management experience. The proposed project effort would clearly not be simple; one wondered how Ames, starting from scratch, would deal with the launch vehicle interface problems, the scientific community, and the challenging data acquisition problems that would have to be solved. Although it was not squarely in my province of lunar and planetary programs, I could see the problems *and* possibilities. I was also aware that Chuck Sonett, an outstandingly good man, was becoming saturated with the papermill aspects of Headquarters and yearned to return to the world of hardware and experiments. Sonett and I paid a visit to Ames, talking with members of the enthusiastic group there, and I also discussed the matter with Ed Cortright and Homer Newell.

The pieces began to come together in May 1962 when Homer Newell, Chuck Sonett, and I met with Smith DeFrance, Director, and John Parsons, his deputy, at Ames. A general approach was outlined, subject, of course, to approval by higher authority. Ames would consider a role in space exploration with a three-part plan consisting of (1) advanced studies and analytical efforts pointed toward a solar probe, (2) project management of an interplanetary program based on the Pioneer series, and (3) establishment of a space science division headed by Sonett, who would be transferred to Ames. The logic for a Pioneer-based flight program included several factors thought to be favorable: the spacecraft concept seemed developed to the point where it was understood; the Delta launch vehicle to be used was proven, and tracking and data acquisition services could be obtained either through the Deep Space Network at JPL or from the Goddard Satellite Net. For starting up a new project and developing the skills of project management, this plan seemed well suited.

After reaching a gentleman's agreement with DeFrance on how the three activities would be viewed by Headquarters and what controls and interfaces would be logical, we also discussed the importance of getting Hugh Dryden's approval, the final prerequisite. On my return to Washington, I outlined for Dryden the general plan we had worked out, and he explained in some detail his concern that in the rush toward space, NASA might inadvertently injure the continuance of the research for which it had become known. But he was sympathetic to the idea, and agreed to consider the proposal on its merits in a face-to-face discussion with DeFrance, if that could be arranged.

In the 1920s, a near-fatal plane crash had caused Smitty DeFrance to pledge to his wife that he would never fly again, a pledge that he honored into the jet age and throughout his directorship of an outstanding aeronautical laboratory. His trips across the country were limited to about one train ride each year. DeFrance made his annual pilgrimage to Washington the following week, endorsed the plan, satisfied Dryden that Ames would continue to excel in research, and Dryden approved. It then took only a few months of countless meetings and memoranda to establish a project office, define the mission, obtain billets for the necessary manpower, arrange funding for the three parts of the plan, and see to Chuck Sonett's transfer and replacement.

As mentioned earlier, this second block of Pioneers was to use the Delta as a launch vehicle. The Delta dictated a modest spacecraft weight of something less than 150 pounds, including instruments. However, since it had been used on many missions, it was thought to be a mature launch vehicle suitable for interface with a new project team. As it happened, the launch vehicle status soon became fuzzy: improvements being made on the Delta for other projects became options for Pioneer, and the new project team became entangled in resolving these choices. With the scientific payload restricted to 20 to 40 pounds, various specific objectives shaped the spacecraft's design. Among these were the desirability of pointing instruments in all directions along the plane of the ecliptic; continuous data sampling from instruments, as opposed to recording and transfer part-time; high data rate transmission from spacecraft to Earth; several commandable modes of operation, allowing experiments to modify their use of the instruments over a period of time; a favorable spacecraft environment, particularly a low residual magnetic field (spurious fields had plagued many prior experiments); and a long useful life in orbit of 6 months to a year. Added to these tough engineering requirements was the fact that the spacecraft was to be a spinner. The net effect of these constraints and desirable qualities was to drive the available technology to the limits, placing unexpected demands on the skills of the Ames team.

A spin-stabilized spacecraft had to be sensitively balanced. Every part had to be designed and placed in such a way that it matched something of equal weight and moment on the other side, and all subsystem components had to be chosen with balancing the spacecraft in mind. It was impossible to do this perfectly on the drawing board; only after actual flight hardware was delivered and installed and the craft experimentally spin tested could the last few pounds held back for balance weights be added and adjusted. Allowances had to be made for everything aboard that moved or that had any weight change during flight.

Magnetic cleanliness was especially important if magnetometers—instruments of particular interest in the interplanetary medium—were not to be affected by the spacecraft's own field. This was a difficulty because almost everything dealing with electrical power and metallic structures could affect the spacecraft field. To measure the very small levels of interplanetary fields, the spacecraft's own field had to be as small as possible, and furthermore, it had to remain the same throughout the mission. Twisted wire pairs, the sedulous avoidance of any cabling that created a magnetic loop, and extensive use of nonmagnetic materials in components all helped. The onboard transmitters used traveling-wave tubes that seemed at first an uncorrectable source of magnetic contamination; the remedy was to spot nearby small permanent magnets oriented to cancel out the tubes' magnetic influences. Considerable effort went into the design of a facility to test the magnetic cleanliness of the spacecraft, not merely at one instant, but under all conditions. This attention to magnetic cleanliness and ways to achieve it were major contributions of Ames and its contractors.

The Franklin array antenna was another concept that had not been extended as far in a technological sense as Pioneer required. This involved not only orienting the antenna on the spin axis but also a design to produce as high a gain as feasible in the toroidal (doughnut-shaped) pattern it produced. As the gain increased, the sensitivity to exact alignment increased; thus the pointing of the antenna had to be corrected as the spacecraft traveled farther away from Earth. For Pioneers it was decided that the spin axis of the spacecraft should be changed as needed by the commanded firing of a small thruster on a boom at right angles to the axis, changing the spin axis and the swath swept by the instruments to the precise plane desired. (It also set up a modest wobble in the spin, like the wobble in a slowing top, but a wobble damper took care of that.) Two different spin-correcting maneuvers were called for: an automatic one during the launch sequence, occurring right after injection, to ensure that the spacecraft's spin axis was as intended; and a commandable one to be initiated from Earth as needed after weeks or months of cruise had altered the geometry of the antenna and instruments. Persons responsive to the aesthetics of mechanism will find pleasure in studying the axis-torquing systems aboard these Pioneers; they were simple, clever, imaginative, *and they worked*!

A communications development highly important to the success of these Pioneers, though not first used on them, was phase-lock operation, a method that allowed the matching of signals from Earth and from the spacecraft to increase the sensitivity of reception over immense distances. In simplified form it worked this way. Let us suppose that a Pioneer is sending its Doppler tracking signals Earthward as it cruises along 100 million miles away. The spacecraft is operating on its own, with its transmitter frequency governed by its own crystal-controlled oscillator. This is a "noncoherent" mode of operation. Simply by listening to it, Earth can manage one-way Doppler tracking of limited accuracy. When the Deep Space Network picks up the weak signal and "locks" onto it, matters take a turn for the better.

Locking consists of directing the signal through a feedback loop and a voltage control oscillator and retransmitting it back at precisely the frequency received from the spacecraft but with a 90° change in phase. In effect, the feedback circuit forces the ground transmitter to match the spacecraft carrier frequency exactly. Once downlink lock is established, the ground transmitter sends its own carrier toward the spacecraft. When this is received, the spacecraft oscillator is automatically disconnected and switched to a voltage control oscillator that generates a signal having a precise ratio to the frequency received from the Earth station. This creates uplink lock, and the two have now formed a coherent roundtrip relationship between spacecraft and Earth that supplies Doppler tracking of exceptional precision. When tracking of this high accuracy is no longer needed, the coherent mode is simply broken at the ground transmitter, and the spacecraft automatically returns to the frequency established by the onboard crystal-controlled oscillator. Two-way phase lock has the particular merit of eliminating the effect of slight frequency drift that may have occurred onboard the spacecraft as the result of temperature changes, radiation, and aging. Another advantage is its ability to supplement the distant, relatively weak and unattended spacecraft equipment with powerful and fresh electronic gear on the ground. It makes possible those astonishingly precise calculations of spacecraft speed and position that surprise nontechnical onlookers.

There were four Pioneers in the block launched from 1965 through 1968, all productive, hardworking spacecraft, informative about the interplanetary medium away from the disturbing influence of Earth. They told us much about the solar wind and the fluctuating bursts of cosmic radiation

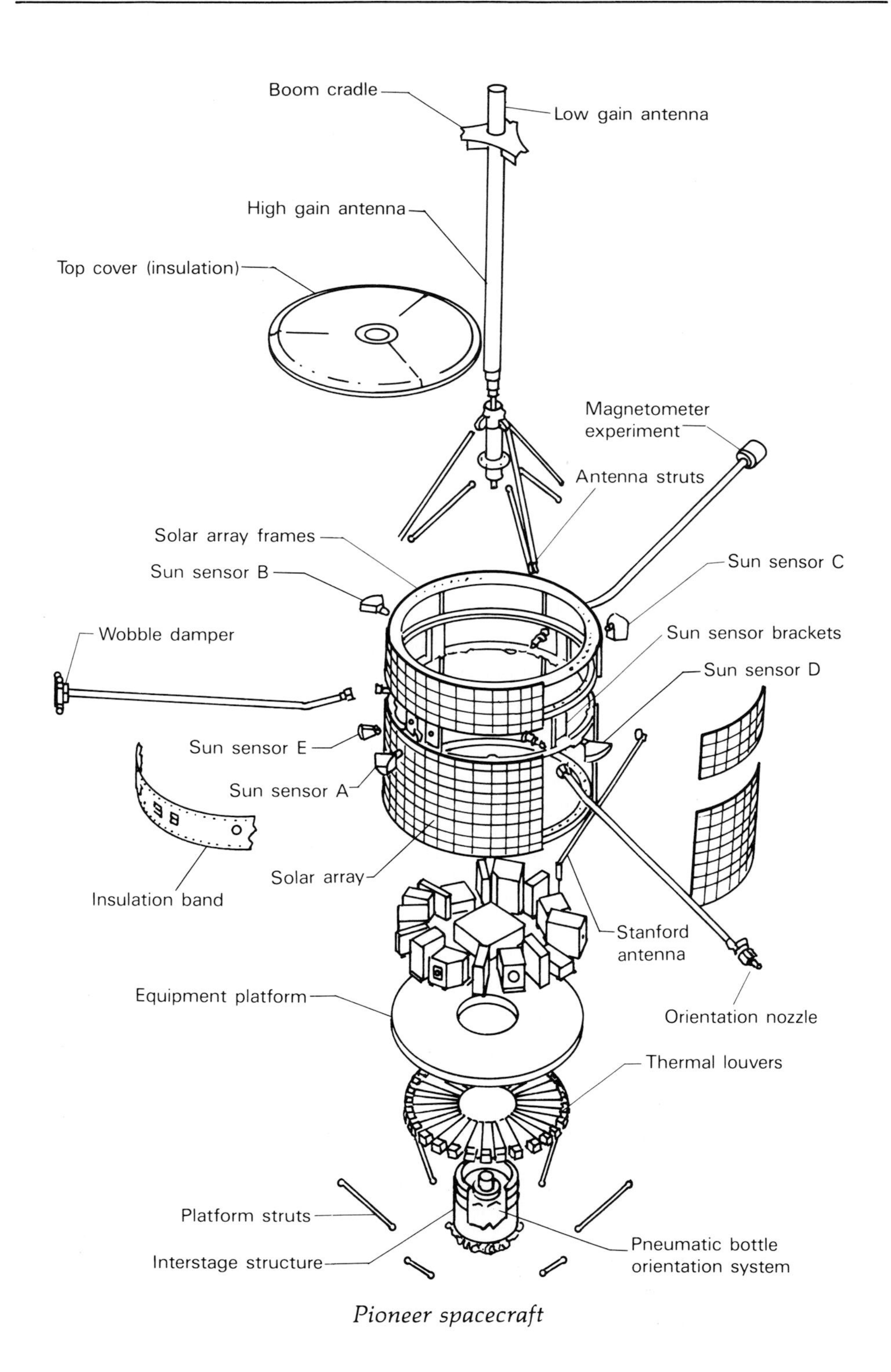

Pioneer spacecraft

of both solar and galactic origin. They traveled in orbits approximating Earth's—two were slightly inside Earth's track and two were outside—and were spaced around the Sun to allow differential timing of the arrival of specific solar events. These four lonely sentinels in space were also an important part of a warning system designed to protect Apollo astronauts against potentially dangerous radiation resulting from solar eruptions.

The original target lifetime of a year in orbit was easily achieved. Nineteen years after the first of the four was launched, all are still working to some degree. Pioneers 6 and 9 still possess all their faculties and still speak when spoken to; Pioneers 7 and 8 have lost their Sun sensors and can respond only when the geometry of their orbits points their antennas Earthward. Such dogged longevity continues to surprise the engineers who worked on them.

Heartened by these quiet successes, Ames began developing a pair of newer, larger, more capable Pioneers designed to attempt more difficult feats. Essentially all previous interplanetary exploration had been directed toward Venus and Mars, Earth's nearest neighbors; now it was time to try to send probes through the unknown barrier of the asteroid belt to scout the distant gas giant, Jupiter. If that could be managed, it might even be possible to make a close pass through Jupiter's unknown radiation belts and gain enough swing-by energy to travel even further, to the ringed planet Saturn.

To suit the requirements of so ambitious a voyage, the spacecraft would have to be drastically modified. At Jupiter and beyond, the Sun would be too distant to create enough solar cell power; the spacecraft would have to carry a radioactive thermoelectric generator, which uses plutonium isotopes to heat an array of thermocouples. The Franklin antenna with its pancake pattern could not produce a signal strength that could cope with such a distance. It would be replaced with a parabolic antenna mounted on the spin axis and aimed back at Earth with rifle-like precision. In place of the earlier Pioneers' simple little thruster systems for initial orientation and another for nudges to precess the spin axis, there would now be no less than four pairs of thrusters arranged so that they could increase or decrease the spin rate, torque the spin axis around in different directions, or even accelerate or decelerate the whole spacecraft. Only one change was not in the direction of bigger and more; the earlier Pioneers had spun at the rate of 60 rpm; the new, larger ones had moments of inertia to hold orientation at a stately 4.8 rpm.

The greater diameter—limited by the fact that the antenna had to fit within the 10-foot shroud of the Centaur second stage—did not ease the lot of spacecraft and instrument designers. At first it was hoped that enough weight could be spared to make these Pioneers partly autonomous, with onboard computers and memory to permit stored sequences of commands. However, as the inevitable weight crunch grew, it became necessary to leave the sophisticated brains on Earth. The long communications time imposed extra stresses on terrestrial controllers. Even though radio commands travel at 186 000 miles a second, the distances were such that it took 92 minutes between command and acknowledgment at Jupiter and more than 170 minutes at Saturn. One of the mind-stretchers of interplanetary exploration is to try to visualize long trains of commands racing at almost unimaginable speed in one direction, and long trains of data and imagery racing back to Earth, both trains, for all their velocity, requiring long periods of time to make the trip.

Fortune smiled on Pioneers 10 and 11, for both proved to be singularly effective spacecraft that accomplished historic missions. Launched on March 2, 1972, Pioneer 10 accelerated for 17 minutes atop its hydrogen-fueled Centaur to a speed of 32 114 miles per hour—at that time, the highest velocity ever achieved by a manmade object. In 11 hours it crossed the Moon's orbit, a distance that had taken Apollo astronauts some 3½ days to traverse. Five and a half months later, past the orbit of Mars, it entered the asteroid belt, an utterly unknown band of scattered subplanetary debris, and in February 1973 it emerged unscathed.

Choosing the best flyby trajectory of Jupiter was agonizing, requiring not just thought about lighting, satellite position, and command sequencing, but also prudent estimation about how close the spacecraft should pass to the intense and potentially disabling radiation known to encircle the giant planet. Complexities arose from the fact that the radiation could generate false commands, and the communications delay could prevent their timely correction. The remedy was to prepare and transmit a series of redundant corrective commands against the chance that false commands would be set off by the intense radiation. Bathed in this steadying electronic reassurance from Earth, Pioneer 10 flew close to Jupiter on December 3, 1973. It was accelerated to a velocity of 82 000 miles per hour by the mass of the huge planet and flung on a course that has taken it out of the solar system. In June 1983 it passed the orbits of Neptune and Pluto, still turning in its stately fashion and responding to questions at a range beyond 2.8 billion miles from the Sun. It is headed in the direction of the constellation Taurus and should reach the distance of the star Ross 248 in about 32 000 years.

Pioneer 10's list of firsts is too long to cover in detail, but it should be credited as the first to fly beyond Mars, the first through the asteroid belt, the first to fly by Jupiter, and the first to leave our solar system. Engineers hope it will be possible to keep in touch until 1994, when Pioneer's radioisotope thermoelectric generators should expire.

Although this was a tough act to follow, Pioneer 11 succeeded and in one important aspect did even better. When it arrived at Jupiter in late 1974, its controllers were better informed about the lethal radiation and were able to manage a closer pass. In addition, the prevailing planetary configuration allowed Pioneer 11 to be guided on a course that flung it off to pass, almost 5 years later, the ringed planet Saturn, never before observed from space. It is a commentary on the pace of planetary exploration in those giddy years that, though the Pioneers added immeasurably to our scant store of knowledge about the outer solar system, the data and images they returned were soon to be overshadowed by more sophisticated exploring machines.

Like its brother, Pioneer 11 is destined to leave the solar system forever, but in an approximately opposite direction. At this writing it is perking along at a range of about 12 astronomical units (over a billion miles) from the Sun, healthy and mannerly. It bears a plate engraved with symbols and mathematical notation telling where it came from and when. This Earth's signature, or builder's mark, is situated in a place that should be shielded for incalculable ages from erosion by interstellar dust. Perhaps somewhere a hundred thousand years from now Pioneer's strange message from Earth will become a haunting reminder of beings reaching out.

Interplanetary Billiards

In the preparation of this book, I was continually reminded of the amazing way the threads of technology were woven into the fabric of missions to the Moon and planets. Developments pursued independently in laboratories throughout the world evolved miraculously to make space flight possible in the 1960s. Some of these developments had been ready and waiting for years; others barely arrived ahead of the need.

Many who contributed to successful space missions were not present to see their research pay off. Because of their special contributions, I would like to illustrate briefly how the works of two such researchers came into play long after their deaths.

The first is Johannes Kepler, who, after working for 18 years, developed laws of planetary motion which are beautifully simple, but crucial to defining the behavior of planetary bodies and spacecraft in orbits. As is usually the case in science, his work built on the efforts of others; his findings would not have been possible without the observations, the positional sightings, the recording of times, and the determination of planetary periods obtained earlier by Tycho Brahe and other astronomers. But by 1618, with these data and his own efforts, Kepler was able to arrive at three basic laws that clearly define the orbital relationships of satellite systems.

His first profound determination was that the planets move in elliptical orbits, with the Sun at one focus. Before Kepler, most astronomers believed that the heavenly bodies moved in circles, and their planetary systems concepts were based on this premise. Of course, the ellipse is a conic section which becomes a circle when its eccentricity is reduced to zero; in other words, a circle is simply an ellipse having coincident focii. Kepler's determination of this feature of planetary behavior was based on years of studying the orbit of Mars, leading to his final conclusion that its path was accurately defined as an ellipse.

His second law, called "the law of areas," says that the line joining the Sun and a planet sweeps out equal areas in equal times. While this simple geometrical relationship was also the result of observations, it provided a basis for relating the speed of a planet to its position in orbit.

The third law, called "the harmonic law," simply says that the square of the time of revolution (in years) of any planet is equal to the cube of its mean distance from the Sun (in astronomical units). While the law of areas enabled changes in a planet's orbital speed to be calculated, from the harmonic law we can obtain either the distance of the satellite from the parent body or the period of revolution, provided the other is known from observation.

At the time these discoveries were presented, Kepler was working in Prague, Czechoslovakia. Many miles away in England and some 47 years later, Isaac Newton studied the laws produced by Kepler's observations, trying to understand the causes. His studies led him to develop gravitational theories as to why the planets move in elliptical orbits, and he produced the mathematical relationships of attractive forces between bodies. I find it interesting that it was almost 20 years before his works were published, reportedly because he was persuaded to do so by Halley, an astronomer who saw the significance of the work. Newton's brilliant discussion in 1687, called the *Principia—System of the World*, showed mathematical relationships for all the known motions of the Moon, the planets, and comets—even the rise and fall of the ocean tides—allowing precise calculations in terms of his laws of motion and gravitation.

Of course these principles are fundamental to all spacecraft trajectory determinations. When Apollo 13 was disabled by an explosion on its way to the Moon, for example, the only hope for recovery required a combination of velocity and lunar flyby distance such that the gravitational effect of the Moon would return the spacecraft to Earth in the proper direction for reentry into the atmosphere. Thus, gravitational forces and their effects, as originally worked out by Kepler and Newton, became the tools by which Apollo mission controllers and astronauts were able to direct the damaged spacecraft to a safe return and recovery on Earth.

The same classical developments, while used in the conduct of every space mission, became strikingly significant in some of the planetary flyby missions. A notable case was the Mariner 10 mission, in which a single spacecraft was sent from Earth to Venus and from Venus to Mercury, making orbits around the Sun and returning to Mercury for three close encounters before completing its mission. It is doubtful that Kepler and Newton ever dreamed of the Apollo 13 and Mariner 10 applications of their findings, but surely they would have respected the engineers who so deftly applied them.

Feasibility studies for using gravity-assist trajectories, as they became known, were first recorded in the 1920s and 1930s, although it was during the 1960s that a team of JPL engineers dedicated to trajectory design became aware of their potential for the Venus-Mercury mission. One of the first studies was a flyby mission to Venus and return to Earth, a trajectory which would be extremely important to a manned flight for reconnaissance. During these studies, it was determined that a near minimum energy condition would exist for a launch past Venus and on to Mercury in 1973.

Soon thereafter, during discussions with the Space Science Board of the National Academy of Sciences, strong endorsement for such a mission was obtained. At the time, the NASA budget was beginning to decrease significantly, funds for future planetary missions were being sharply curtailed, and the Venus-Mercury Mariner concept of using interplanetary forces to obtain more science per dollar was exciting. Project evolution and mission operations provided some of the most memorable planetary experiences to date.

Mariner 10 embodied a combination of many advanced technologies that had evolved over the years. The gravity-assist concept for redirecting the orbit of a spacecraft, while requiring no additional rocket energy, demanded extremely accurate guidance and control systems to produce the precise flyby distances and velocities necessary. Earlier missions had refined our knowledge of the factors that tend to affect orbits, such as gravitational fields for the planets, and solar radiation pressure. Advances in attitude control and autopilot systems, plus improvements in tracking, allowed precise determinations of the trajectories and initial conditions required for velocity corrections. Added to this were improved vernier rocket systems used for trajectory adjustments.

To be able to achieve the close flyby of Venus with precision would require multiple trajectory corrections—at least two between Earth and Venus and two more between Venus and Mercury. You will recall that during the first Mariner mission to Venus it was debated whether a midcourse maneuver should be tried because of the hazards and uncertainties associated with the remote spacecraft orientation maneuver and rocket firing. These technologies had advanced such that we were confident the spacecraft could be put within 250 miles of an aim point at Venus, so that after passage by Venus it would approach Mercury with enough precision to obtain meaningful scientific data.

Another technology that had literally soared during previous years was communications. Recalling that the first Mariner mission to Venus had a data return rate of 8⅓ bits per second, Mariner 10 transmitted up to 118 000 bits per second—making it possible to send TV data in virtually real time while concurrently transmitting other science and engineering data at 2 450 bits per second. These phenomenal increases in data rates, plus a new command and control system for processing and programming, provided 21 data modes for television or nonimaging science, engineering data, and data storage playback.

Also significant to the Venus-Mercury spacecraft was the fact that the solar constant increased in value by four and a half times during Mariner's trip to the vicinity of Mercury, which orbits close around the Sun. Thus the thermal control system for the spacecraft could not be passive, but had to incorporate features such as solar panels that could be rotated "edge-on" toward the Sun to help keep temperatures within bounds. Mariner 10 used a combination of sunshade, louvers, and protective thermal blankets to "keep cool" during the close approach to the Sun.

One of the solar protective devices was an umbrella-like sunshade made of a Teflon-coated glass fiber fabric known as beta cloth. This simple device, suggested by Robert Kramer of NASA Headquarters, unfurled in the same manner as an umbrella, and shadow shielded the rocket system and parts of the spacecraft when pointed toward the Sun. Although Mariner 10 experienced temperatures near the Sun as high as 369° F—hot enough to melt tin, lead, even zinc—the temperature of its solar cells never exceeded 239° F. The temperatures of the television cameras dropped so low at one time during the flight that there was concern that the quality of the pictures might be degraded, but this did not happen.

In addition to the challenge of being the first mission planned for a two-planet encounter, Mariner 10 faced a number of obstacles in its approval phase that almost kept it from being. As already mentioned, when the project was presented to Congress in 1969, support for the space program had begun to wane and reductions in scientifically oriented projects were the norm rather than the exception. The Subcommittee on Science and Astronautics headed by Joseph Karth was giving a great deal of emphasis to space applications and putting pressure on NASA to use space for practical purposes. This put proposals such as the Venus-Mercury mission into direct competition for funding with Earth resources, communications, and other applications missions because they were all considered by the same congressional subcommittee. It is in the record that the chairman told John Naugle, Associate Administrator for Space Science and Applications, that he didn't believe he was giving enough priority to applications as opposed to science, and that he was going to withhold funding for the Venus-Mercury mission until priorities were changed. The House of Representatives did not authorize the mission at first, and it took persuasion from the Senate plus a conference between the two houses to obtain fiscal year 1970 funding.

Even after Mariner 10 was put in the budget, NASA Headquarters officials were concerned that it would survive only if costs were kept low. According to Bob Kramer, who was then Director of Planetary Programs, estimates based on past Mariner experience showed that the job would cost about $140 million. JPL desperately wanted to do the mission, and Bill Pickering sent a letter to Headquarters saying, "I will absolutely guarantee that JPL will do the job for 98 million dollars." This strong guarantee by the director of the project center was encouraging and was accepted by NASA Headquarters with some trepidation.

Bob Kramer told me that the budget allowed only about $6 million for the video-imaging system, including the camera, all associated mission costs, data analysis, and photographic prints. The head of the imaging team was a comparatively young scientist from CalTech named Bruce Murray. Bruce knew what the budget meant, but, being an aggressive person, he also believed that JPL might be able to find a way to modify the budgeted amount. After the mission was approved, Bruce came to Washington and pointed out that the spacecraft would be going past Mercury faster than any spacecraft had ever flown by a planet before—something like three times faster than any previous planetary encounter. At that speed, he said, the cameras would not really see anything; they would produce only a blur. He proposed a film system with image motion compensation patterned after the system on Lunar Orbiter which developed film in flight and read it back slowly. A cost estimate was made for such a system, and, according to Kramer's memory, it was something like $57 million. So he said, "Bruce, that won't quite fit into your six million dollar budget."

Not giving up easily, Bruce came back with a proposal for a dielectric camera, being developed by RCA, at a price estimate of about $40 million, assuming that its development was successful. Kramer told Bruce that $6 million was still the budget and that such a camera wouldn't fit.

Finally Murray started talking to the systems engineers and the imaging science team about the problems. They all recognized that the low transmitter bit rate was a major factor, along with limited tape recorder capability. As Murray had pointed out, a physical aspect of the high-speed flyby was the fact that high-resolution data had to be obtained very quickly; there seemed to be no way that pictures could be taken and stored satisfactorily for later transmission.

The imaging team and the systems engineers began to collaborate on the development of a concept for sending back a quarter of the pixels for each image in real time, while storing the others for transmittal later. This would make the best use of the limited communications and recording capabilities and ensure that some picture data were obtained, even if recording and later transmission did not pan out.

After the telecommunication, camera, and recorder tradeoffs were studied, the same vidicons that had been used successfully on the 1971 Mariner orbiter were adopted, with the basic lens or optical elements extended to 1500-millimeter focal lengths so that they became real telescopes, able to provide magnified images. The option of sending back only one-quarter of the full frame (every fourth pixel) or the full frame was retained. Either mode could be commanded from Earth. In the quarter-frame mode, thousands of images could be sent that were suitable for mosaicking the whole planet. By scanning across the planet during the fast encounter, it was possible to obtain excellent photographic coverage.

Although the system was designed to provide good coverage at high resolution, one desire was not fulfilled: full-frame imaging of the planet. For 1971 Mariners, one camera had a wide-angle lens and one had a narrow-angle lens, so that both types of information could be obtained. Since this option would not fit within the $6 million budget, ingenuity came into play again. Small mirrors were added to filter wheels used for viewing in different colors, so that images could be directed toward small wide-angle lenses mounted on top of the cameras. The mirrors and simple lens systems (just 3 or 4 inches long and 1½ inches in diameter) allowed each instrument to become a wide-angle camera by simply flipping the filter wheel around to the mirror. Thus, for $6 million the imaging team got almost everything it wanted, ranging from extremely high-resolution images of Venus and Mercury to wide-angle views of the planets on approach and departure.

Such ingenuity helped ensure that the entire Venus-Mercury Mariner project was completed within the $98 million guarantee. The outstanding project management effort was led by Eugene Giberson, the first manager of

Surveyor who was replaced when the project got so deeply in trouble. The success of Mariner 10 provided proof that Gene really had what it took to be a good project manager, and I was very proud of his comeback. This was not to be his last success; he also managed the Seasat mission which taught us much about Earth's oceans and was the forerunner of many new activities.

With the remarkable "parentage" provided by the team of scientists and engineers who devised the Venus-Mercury Mariner mission, it is not surprising that Mariner 10 developed a very interesting personality of its own. The mission became one of the most exciting to follow on a day-by-day basis, as troubles developed and were overcome in unexpected ways.

Within a day after launch, when instruments were being checked out, the camera heaters would not come on. Heaters were needed to keep the vidicons at reasonable temperature when the craft was far from Mercury; it was easier to shield the spacecraft when it was near the Sun and to provide heat when it was far from the Sun. There was great concern that the camera optics would cool down so much that they would not remain in focus, so the vidicons were switched on to maintain some heat within the cameras. With these precautions, the temperature of the cameras stabilized at low but viable values, and the picture data never showed any degradation as a result of the low temperatures.

About 2 weeks before encounter with Venus, the heaters for the TV cameras mysteriously came on. There had been great concern that the cameras might not operate properly during the Venus encounter, as their temperature had dropped well below freezing. It was not possible to know exactly what had happened, but engineers decided the problem might have been the result of a short circuit in another heater which had been biasing the TV heater. To avoid any risk to the camera heaters, the heaters in the related, suspect circuit were turned off. By this time, the spacecraft had warmed up enough with its closer approach to the Sun that not all of the heaters were needed.

Two months after launch, the most significant power-related problem occurred when the spacecraft automatically switched from its main to its standby power mode. This automatic switchover was irreversible. It was of great concern because of the possibility that it might have been caused by a fault common to both power circuits and might eventually cause the backup power supply to fail, ending the mission prematurely. Following the automatic switchover to the backup system, engineers were very careful when making changes in the power status of the spacecraft. Care was also taken in maneuvering relative to the Sun to avoid automatic switchover from solar panel to battery power, should sunlight be lost.

Another problem occurred on Christmas Day when a part of the feed system of the high-gain antenna failed, causing a drop in signal strength. Diagnostic commands provided indications that temperature changes during flight may have caused the problem. It was of concern because should the high-gain antenna not perform properly, the real-time TV sequences would not be possible at Mercury encounter, greatly reducing the coverage and the benefits from the clever mosaic technique that had been worked out. About 4 days later, the feed system healed itself, and the high-gain antenna performed normally again. However, relief was short-lived, for within about 4 hours the fault reappeared, indicating that it was an intermittent glitch which might recur at any time. The problem with the antenna was a threat throughout the mission, but it apparently was solved by the increase in temperature and did not compromise any of the pictures.

A serious attitude control problem developed about a week before the flyby of Venus. The trouble occurred after Mariner 10 started a series of eight calibration rolls to allow the scan platform to obtain diffuse ultraviolet data over wide regions of the sky. Oscillations began suddenly in the roll channel of the attitude control system, causing the expulsion of attitude control gas at a very high rate. Watching the gas pressure drop, mission controllers knew that the spacecraft would die if this continued. In the hour it took to recognize, analyze, and respond to the problem, about 16 percent of the 6 pounds of nitrogen gas—the total supply of attitude control gas—had been lost. When the fault was determined to be the result of a gyro-induced instability, the gyros were turned off and the gas loss stopped. Later it was decided that the oscillation was caused by a long boom supporting a magnetometer at some distance from the spacecraft, which apparently entered into a resonant dynamic relationship with the attitude control system. After an extensive analysis, commands were sent to place the movable solar panels and scan platform in such a position that solar pressure could help prevent the oscillation and avoid further loss of gas. Spacecraft attitude maneuvers and trajectory corrections were also modified to minimize gas usage.

It had been planned to use the gyros during the Venus encounter to ensure proper stabilization of the spacecraft. The reason was that the Canopus tracker, a light sensor, might be affected by particles near the spacecraft, by the background light from the planet, or by some other source which could cause a loss in attitude stabilization at a critical period during encounter. As there was not enough time to determine the cause of the gas loss problem, a quick decision was made to take these risks and maintain attitude control during flyby using the celestial references of the Sun and Canopus. Everything worked beautifully during encounter, and all the data, including a grand total of 4165 images of the cloud-shrouded planet, were obtained as planned.

Once past the successful encounter with Venus, engineers had to decide how to plan the correction manuever that would allow the spacecraft to go past Mercury without using the gyros. Experiments were performed with the tilt of the solar panels to determine how to use these as "solar sails" or "rudders" and thereby save attitude control gas. About 2 weeks after encounter with Venus, the gyros were tested again. They seemed to function correctly at first, but then the oscillations began. As a result, a decision was made to plan a trajectory correction with a so-called "Sun-line maneuver." This required a wait until the spacecraft attitude relative to its trajectory was such that a simple firing of the rocket without attitude change would produce a suitable trajectory correction. Calculations showed that this would delay arrival at Mercury by 17 minutes, but would still be satisfactory for the science objectives.

Shortly after this decision, the spacecraft lost its Canopus reference and began drifting about the roll axis; the gyro mode had to be turned on and off to stop the motion and to reacquire Canopus, resulting in additional loss of the precious attitude control gas. Similar events were to occur about 10 times a week through early March, when conditions were right to make the Sun-line course correction.

With the particular orientation of the spacecraft for this maneuver, it was not possible to obtain good Doppler data during the rocket firing; a considerable amount of tracking was needed after the maneuver to determine whether it had been successful. Refined trajectory calculations finally showed that the spacecraft would be passing 124 miles closer to Mercury than had been planned, but still within a satisfactory range. Like an unruly child who behaves very badly and becomes a model child just as anxious parents expect to be embarrassed, Mariner 10 began to function perfectly again just prior to its encounter with Mercury. The high-gain antenna had recovered, never to fail again, and high-resolution photographic coverage of Mercury was achieved as planned. This first return of high-resolution photographs of Mercury produced exciting new information of a Moon-like planet, with many features that had never been seen from Earth. Techniques developed for the mosaic process worked as planned, and the wide-angle lens feature worked well. Scientists everywhere were ecstatic.

Shortly after encounter, in now typical Mariner 10 fashion, problems began to recur. An additional 90-watt load was registered on the power system, accompanied by a rapid rise in the temperature of the power electronics bay. This anomaly, following the still unexplained switchover from primary to standby power early in the flight, was indeed foreboding. Many tired engineers spent hours developing similies to the problem and devising work-arounds to control the temperature in the best possible way without adding stress to the power system. Other failures followed during the same week: the tape recorder power turned on and off several times without command and soon failed altogether; commands to change the transmit power level were not acted upon; and the flight data subsystem experienced a failure which caused a dropout of several engineering data channels, making it very difficult to determine what was happening and to nurse the ailing spacecraft around the Sun to reencounter Mercury. Since analyses of the loss of attitude control gas showed that gas usage would have to be reduced well below the normal cruise rate if the spacecraft were to encounter Mercury a second and third time as hoped, further multiple trajectory correction maneuvers had to be conducted, and some way had to be found to use the gyros without causing the oscillation problem. Engineers had by this time determined how the movable solar panels and the high-gain antenna worked as "solar sails," so that attitude control could be maintained, and some slight modifications in the trajectory could be effected using solar pressure.

To redirect the spacecraft for a return to Mercury, a very large maneuver was required which would have meant a long rocket burn. To prevent overheating of the rocket engine, the maneuver was programmed in two phases. This two-phase maneuver refined the aiming point of the spacecraft so that it would return to the vicinity of Mercury after making a pass around the Sun. As the spacecraft passed behind the Sun from Earth, data were obtained on the Sun's corona, adding to the planetary data collected about Venus and Mercury.

When the fifth trajectory correction maneuver was made in July 1974, the spacecraft was on the far side of the Sun from Earth. Just after the spacecraft began its attitude change for the maneuver, all the pens on the plotters dropped to zero and made straight lines, indicating that telemetry signals had ceased and no data were being returned. In spite of the fact that the mission controllers were not able to see what happened, the spacecraft completed its automatic commands exactly as they had been stored; after the maneuver, the spacecraft commanded itself back to the cruise orientation, and telemetered signals were again received. With this new orbit, a passage by Mercury for the second time was assured; in addition, the trajectory change caused by the encounter with the planet and its gravitational field made possible a third encounter after another orbit around the Sun.

Following a brilliant performance in the vicinity of Mercury, Mariner once again acted up. This time it lost Canopus lock and began an uncontrollable roll. The automatic reaquisition sequence had been inhibited to save gas, and repeated reacquisition attempts using commands on the basis of the star tracker roll-error signal telemetry were unsuccessful. Each of these attempts required a momentary turn on of the gyros and the attendant use of the almost depleted gas supply. Roll axis stabilization had to be abandoned for this portion of the trajectory in which the attitude, other than solar orientation, was not critical. A roll-drift mode, allowing the spacecraft to roll slowly, was used. The rate was controlled by differentially tilting the solar panels; in a sense these became "propeller blades," with pitch changes commanded from Earth to moderate the roll rates.

This complicated operational technique was made more difficult by the loss of the engineering telemetry channel that had occurred earlier. But, after some study, engineers found that they were able to measure the roll rate by analyzing the signal from the low-gain antenna. This signal varied with roll position due to the nonuniformity of the antenna radiation pattern. Of course, signal strengths had been measured during testing before the mission began; after a few hours this technique became a suitable means of determining the roll attitude and drift rate of the spacecraft. By using this "roll stabilized" mode, only 25 percent of the normal cruise usage of attitude gas was required, allowing Mariner to reach Mercury for the third time with a slim margin—just enough to cover the encounter and a few days after. Three important trajectory correction maneuvers were completed, and the spacecraft was placed on a very close planetary encounter, determined to be only 2035 miles above the surface.

A few days after the encounter, trouble again developed, and the final significant drama for Mariner 10 engineering operations occurred. During an attempt to reacquire Canopus, the spacecraft rolled into a position such that the low-gain antenna was in a deep null and communications with Earth using the 85-foot dishes were completely interrupted. To compound the problem, the large dishes of the Deep Space Network were tied up with a very important Helios mission that was approaching the Sun. In order to save Mariner, the controllers of Helios were asked to allow some use of the 210-foot antennas, and they acceded to this request. Using the more powerful transmitter and antenna at Madrid, it was possible to arouse the Mariner spacecraft and command it to its proper orientation just in time for the third flyby of Mercury. The third encounter produced some of the most remarkably detailed pictures of the planet and additional information on the magnetic field because of the very close 620-mile flyby trajectory.

About a week later, Mariner 10's nitrogen supply was depleted, and the spacecraft began an unprogrammed pitch turn which told engineers it had finally exhausted its capabilities. Commands were immediately set to turn off its transmitter, and radio signals to Earth ceased. It then became a silent partner to Mercury, forever in orbit about the Sun. But it had performed brilliantly, and all associated with it had learned to respect its personality. However obstreperous, Mariner 10 always came through in the crises.

The interplanetary billiards successfully initiated by Mariner 10 and used by Pioneer 11 to swing by Jupiter and on to Saturn were followed by the spectacular flights of Voyagers 1 and 2. Both spacecraft have visited Jupiter and Saturn, with close encounters of several moons in orbit about those gas giants. Voyager 1, having completed its planetary exploration, is now sailing into the far reaches of the solar system. Voyager 2 is on a course to Uranus and is expected to continue to Neptune for close encounters in 1986 and 1989, respectively.

It is appropriate to class the Voyagers as planetary *systems* explorers, for, by judicious use of sophisticated navigation and guidance techniques, they examined 20 known satellites and more than a dozen new ones discovered during Pioneer and Voyager missions. The four planet-like Galilean satellites of Jupiter were of special interest, as was Titan, the almost Earth-like moon of Saturn. The Voyagers also examined Saturn's six icy satellites, of interest because water-ice is the dominant material on their surfaces. Among the most exciting findings about the moons of Jupiter and Saturn is the fact that several of them are still active volcanically; some have active atmospheres, and Titan at least may have oceans of liquid nitrogen or methane.

These extraordinary achievements resulted from a fall-back position taken after a program called "The Grand Tour" failed to win approval. At the time gravity-assisted trajectories were being studied for the Venus-Mercury mission, engineers discovered that in the late 1970s the outer planets would be roughly aligned in a manner to make all of them observable by a single flyby spacecraft. After passing by Jupiter, the craft could be redirected toward Saturn; from there it could go by Uranus, then Neptune, and finally, after about a decade, past Pluto.

This exciting opportunity had last occurred when Thomas Jefferson was president and would not recur until some 175 years later. To do the opportunity justice, a set of sophisticated and expensive spacecraft would have to be developed, for the requirements of the long-lived, complex operation would be demanding. Those of us supporting the plan believed the potential returns from a single program seemed too good to pass up, but waning interest in space activities and troubles with Viet Nam and other matters made the proposition less attractive to Congress. It was not very long before lesser, more affordable goals were set for a mission to Saturn by way of Jupiter. The Voyager program, approved in 1972, preserved the basic concepts of the multiplanet flyby, using advanced Mariner-class spacecraft that were the most complex ever designed and built by JPL.

These latest operational planetary spacecraft and their marvelous systems can be compared with Mariners or Vikings from the configurational viewpoint; however, a principal difference is the large central antenna 3.7 meters in diameter, outsized because of the communication requirements for the far travelers as they journey to the outer reaches of the solar system. The other obvious configurational differences are the extensible booms. One provides for a steerable platform containing TV cameras and other science instruments; another serves to locate sensitive magnetometers away from the magnetic fields produced by the spacecraft. The same Mariner-like, multisided bus structure was used, but a bank of three radioisotope thermoelectric generators replaced the solar panels which could not provide enough power so far from the Sun.

For such long distance operations, redundant radio systems were employed; even though they were expected to operate up to a billion miles from Earth, the transmitter power for each is only 23 watts. This does not seem like a large gain over the 4½ watts used by Mariner 2 to transmit at 8⅓ bits per second, but the larger antenna and several other advances in technology resulted in a bit rate at Saturn of a whopping 44 800 bits per second. Since the communication system is as critical to an automated exploring machine as a reentry rocket is to a manned vehicle, the tremendous strides in telecommunications technologies deserve great applause.

At the great distances being traversed by Voyagers, accurate position determinations are aided by the use of simultaneous, two-station ranging to increase the viewing baseline. Two Earth stations many miles apart work as a team to obtain angle and Doppler data. Uplink transmissions at S-band frequencies and two downlink frequencies at both S-band and X-band that are coherent with the uplink provide discrimination for the dispersive effects of charged particles along the signal paths.

Maneuvering among the moons of Jupiter and Saturn and flying through the rings of Saturn have been facilitated by optical guidance techniques first experimentally used by Mariners 6 and 7. In principle, a camera mounted on an accurately positioned scan platform can center an object in its field of view and indicate pointing direction relative to spacecraft coordinates. The information from the optical system can be used to adjust the platform toward other objects or to reorient the spacecraft for retromaneuvers, if desired. Changing from inertial coordinates to target object coordinates can improve the approach and flyby accuracies. Optical techniques combined with Doppler systems used for baseline cruise have been very effective in obtaining close-up images of the satellites of Jupiter and Saturn.

From the navigation standpoint, the Voyager 1 encounter with Saturn was probably the most complex ever experienced. Saturn's moon Titan, the largest moon in our solar system, was of special interest for a close flyby. This was a difficult requirement to meet, partly because precise information about Titan's mass and orbit was not available in advance. Several very small rocket burns were used along with optical data to refine the trajectory, and during the Titan encounter Doppler data were processed quickly to allow accurate instrument-pointing adjustments for the outbound imaging of the satellites Mimas, Enceladus, Dione, and Rhea.

Charles E. Kohlhase, Voyager Mission Design Manager, might also be labeled "Chief Navigator." It was Charlie's job to plan the trajectories so that proper flyby times and distances would result in desired velocity changes and viewing geometries for the scientific instruments. Also his was the challenge of determining course correction rocket firings. Because of the relevance of attitude orientation and celestial mechanics, his team was also able to figure out how to rotate the spacecraft for pointing when the scan platform azimuth system malfunctioned.

It is impossible to outline the strides that have been taken during the years since Charlie first began calculating trajectories for guidance and control of Mariners. I can express my respect and admiration for his achievements and for those of his colleagues, but I am sure that even greater acclaim would come from Kepler and Newton, were they here to see their principles being applied.

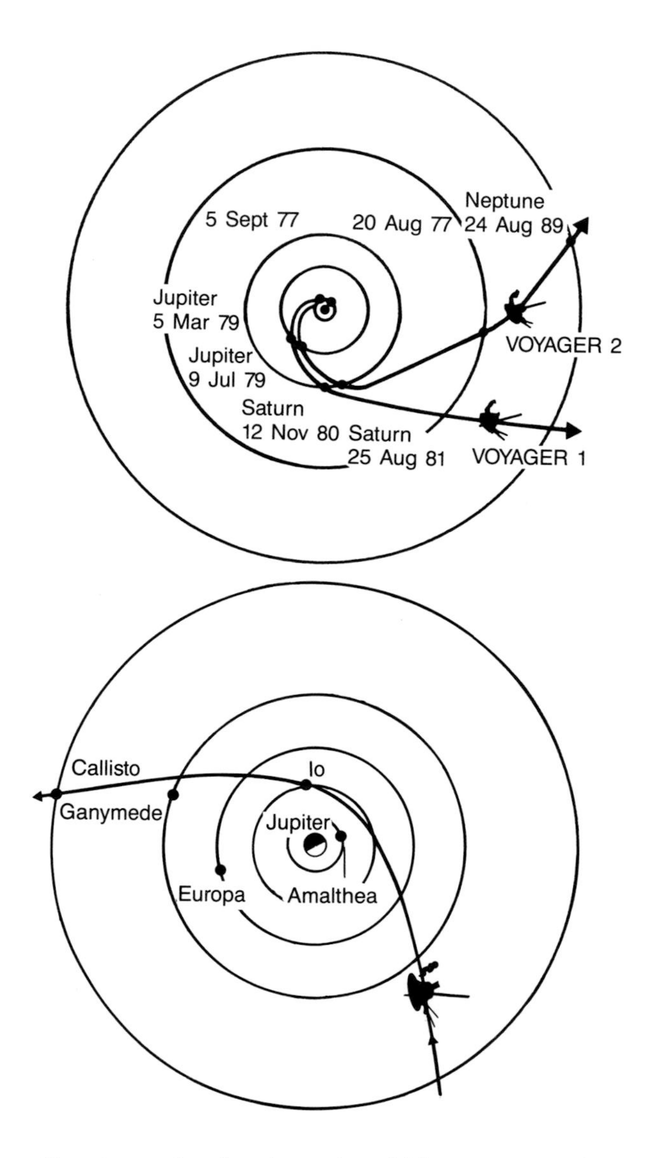

Gravity-assisted trajectories of Voyagers 1 and 2

Partners to Man

Without sanctifying the results through comparison, man's creation of spacecraft "in his own image" follows the example set by God in the creation of man. Only through faith can we hope to assess God's noble reasons for creating man, but we clearly understand why automated spacecraft were devised to help explore the Moon and planets. At the time, they were absolutely essential to do what we wanted to do but were not capable of doing ourselves; our only choice was to devise machines that could go into space in our stead, doing our bidding and performing under our direct control.

It is fitting to note that the evolution of our God-given talents made it possible for us to do the technical things needed for building and operating spacecraft. It is also noteworthy that the spacecraft we have created, while similar to humans in function, do not exhibit the most significant qualities of life. Nevertheless, they have served us well as partners, going forth in the name of our country, seeking answers, and setting precedents for civilized societies now and to follow. While broadening man's horizons, they have also provided a better awareness of our earthly environment and its uniqueness in the universe.

Lest it appear we have overlooked the fact that man himself ventured into space almost at the same time as our robots, I would suggest that the complementary aspects of manned and automated missions typify the trend for the future. Perhaps the inference that automation was devised as a means for exploring until we could go ourselves was true but incomplete—such limited thinking caused an undue polarization of views concerning the merits of manned versus unmanned missions during the first two decades of space exploration. The truth is that there were no such things as unmanned missions; it was merely a question of where man stood to conduct them. In some cases he sent his instruments and equipment into space while controlling them remotely, and in other cases he accompanied them in spacecraft equipped with suitable life support systems. It should be noted that the spurious arguments over the merits of manned versus unmanned missions were never between machines and men, but between men and men. Perhaps the mere fact that arguments ever occurred over the competitive aspects of manned versus automated missions attests to the potential of robotic partners in space operations.

Now that we have a quarter-century of experience to look back on, it is possible to assess some of the results and contributions made by our exploring machines. Perhaps one of the most significant observations, made by a number of writers, is the fact that the adventures of our automated spacecraft have been enjoyed and shared with mankind in a real-time drama, literally unfolding before millions of eyes. While I have often thought of these missions as similar to the expeditions of great explorers like Lewis and Clark, one beautiful consequence of today's communication systems is the immediacy of sharing the experiences, findings, and results of exploratory missions. I have heard a number of scientists express the feeling of "being there" with Mariners, Surveyors, and the like, for they could associate with those lifelike machines, superposing their own human characteristics, without having to consider "the other human being." Thus, our automated explorers have truly been extensions of man in fulfilling our exploratory desires in a briefer span of time and with broader participation than otherwise would have been possible. They have allowed us experiences that in the past would only have been available to the hardy explorer, without our risking life, limb, and personal resources.

In looking back we fondly recall the additional knowledge of Venus provided by Mariner 2. Its flyby trajectory led to an accurate determination of the planet's mass and orbit, fundamental parametric qualities essential to further scientific studies, if not of particular interest to most people. Still, I think that almost everyone shares somewhat in the pride of knowing that we gained better insight into our nearest planetary neighbor on that first mission. Mariner 2 also gave us our first conclusive information about the dense atmosphere of Venus, including a measure of temperatures at different levels in the clouds. While modest findings in themselves, the close-up data from Mariner enhanced the value of sightings from Earth and from scientific studies employing assumptions now better qualified.

Mariner 5, a somewhat improved descendant of Mariner 2, was transformed from a spare spacecraft to perform another flyby visit to Venus in 1967. Equipped with better instruments, it made definitive measurements of Venus' magnetic field, ionosphere, and radiation belts. It told us much more about the composition of the atmosphere and disappointingly concluded that Venus had a chemically polluted environment that was at least 750° F near the surface.

Almost 5 years later, the Soviets succeeded in probing the atmosphere with Venera 7, confirming with in situ measurements that carbon dioxide was the predominant constituent. On its way to Mercury, Mariner 10 swung past Venus for another look in 1974, followed shortly by Veneras 9 and 10, which found the surface of the planet to be firm and rocky.

In 1978 the Mariners and Veneras received help from two Pioneers that orbited the planet and fired probes into its surface. The Venus Pioneer was the first spacecraft to be placed in orbit around Venus, supplying a radar map of the surface. A giant rift canyon, the largest ever discovered in the solar system, measured 15 000 feet deep and 900 miles long. In addition to much improved knowledge of the planetwide cloud coverage, Pioneer Venus 1 detected almost continuous lightning activity. Pioneer Venus 2 launched four entry probes 3 weeks before reaching Venus; during entry these probes relayed direct measurements on the structure and composition of the atmosphere. They also provided temperature and pressure profiles, plus tracking data showing deviations in their trajectories which gave indications of Venusian wind velocities. Since that time, visits to Venus have been left to Veneras.

Mercury was looked over well by Mariner 10 during its two-for-one flyby after leaving Venus, using gravitational assistance to change course and velocity. In addition to its first use of a planetary swing-by trajectory to visit another planet, Mariner 10's three encounters with Mercury were of great interest. This feat surely put Mariner 10 in the lead as our most economical and efficient spacecraft. Being closest of any planet to the Sun, Mercury would not even be "a nice place to visit," so it was just as well that Mariner 10 went there for us. We now have answers to many questions concerning Mercury that also help to complete our understanding of solar system mysteries.

Mars was first visited by Mariner 4 in 1965. Although this first successful Mars mission taught us many things, perhaps the most significant was the finding that Mars' surface is pockmarked with craters, much like the Moon. The second finding of great interest was the fact that its atmosphere is very thin, about one-tenth the density of Earth's. Four years later Mariners 6 and 7 told us quite a bit more about Mars' planetary environment and atmosphere, improving knowledge no doubt helpful to the design of the Soviet Mars 3, which landed in 1971 but ceased transmitting 20 seconds after landing. During the same Mars opportunity Mariner 9 went into orbit while Mars was totally obscured by a gigantic dust storm. After waiting several weeks for the dust to clear, Mariner 9 did a good job of filling us in on details while returning 7300 photographs from orbit. New findings of special note were the huge volcanoes up to 16 miles high and a "grand canyon" over 3000 miles long. Mariner 9 also observed Phobos and Deimos from orbit, giving us our first close-up views of those small satellites of Mars.

Viking orbiters and landers performed so notably while exploring Mars in 1976-83 that, even though their adventures were presented in two earlier chapters, they deserve mention again. The fact that two orbiters, two entry vehicles, and two landers conducted a significant exploratory expedition, sending back detailed information during two Martian years, will be a hard act to follow. Perhaps men will accompany the next robots sent to Mars.

Pioneers 10 and 11 and Voyagers 1 and 2 receive credit for firsthand observations of the asteroid belt, Jupiter, Saturn, and their moons and asteroid rings. The beautiful images returned by these successful craft have not only awed scientists but surely have appealed to the artistic qualities in all who respect the beauty of color and form. There is something inspiring about the giant red spot on Jupiter, more so now that time-lapse images have clearly shown the dynamic nature of this feature. Is there any form more enthralling than beautifully colored Saturn highlighted by its shining, geometrically perfect system of rings? Or is there any more tantalizing object begging scientific examination than Saturn's satellite Titan, the largest moon in our solar system and the only one thought to have oceans and an atmosphere resembling those believed to have existed on Earth during its primitive times? Voyager 1 brought in a flood of findings but is all the more memorable for clearly framing these fantastic questions.

While these summary paragraphs do not do justice to the total achievements of our principal planetary missions, perhaps they will serve to verify our thesis concerning the promise for automated spacecraft and their role as partners to man. None of the missions described could have been done without them, and it may be a long time before we visit all the planets in person—even if we want to.

Since there are powerful arguments for using automated spacecraft to conduct planetary missions, why are there debates among sophisticates regarding the merits for both manned and unmanned missions to the Moon? One reason may be that man, as a *generalized* "scientific instrument" suited to exploration and discovery, has not been equalled by any manmade package fitting the same mold. Another reason may be that some see a competition between the two approaches for dollars, manpower, and prestige, causing them to choose sides whether the competition is real or imagined.

In some ways, it was almost ironic that the technologies needed for manned flight evolved concurrently with those needed for automation. Many of the needs were the same; some were not. Of course both modes shared technologies for rockets, guidance, and trajectory control systems; the biggest differences were the one-way nature of automated missions that obviated the Earth return requirement and the life support needs of human flights into space. I believe the emphasis for capable automated missions rested largely on control and communications technologies, whereas requirements for manned missions depended more on solving reentry and environmental control problems. At any rate, the heritage of missilery served to bring capabilities into focus so that manned craft or automated machines offered options or complementary functions in the same decade.

A strong reason to expect the far-term blend of manned flights and automated missions is that increased capabilities in certain functions can be easily achieved with automated systems. For example, machines can have response times much faster than humans; instrumentation derived from extensions of microscope and telescope technologies are obviously superior to man's naked eyes. On the other hand, man's ability to assimilate input data, to retain and properly integrate information, and to reason, plus a significant array of physical capabilities, make him a powerful machine for which there is no equal.

Setting aside that philosophical discussion, let us return to our review of achievements with a look at the missions to the Moon. Here we can see not only the contributions of automated systems, but we can examine the complementary qualities of manned and automated missions with an eye toward future possibilities. For completeness we must recognize that Luna 1 impacted the Moon carrying a Russian medallion and that Luna 3 returned a low-resolution picture of the far side in 1959. It was 5 years later that Rangers 7, 8, and 9 showed close-up details. In 1966, Luna 9 and Surveyor 1 landed, testing the suitability of the surface and topography for Apollo landings. Since the findings of Surveyor corroborated the engineering model that had been used for designing the landing gear for the Apollo Lunar Module, it might seem that Ranger and Surveyor missions were unnecessary. Perhaps this is so, but no one knew at the time. Suppose this had *not* been the case—how would we have felt today after an Apollo discovery that such a landing could not be made safely?

Even given the acceptability of Surveyor sites, the reconnaissance performed by Lunar Orbiters greatly facilitated the planning and execution of Apollo missions. Such mapping and topographical data as were made available by Lunar Orbiters would have had to be obtained some other way, perhaps at far greater cost and risk. It is also doubtful that the broad area coverage of the Moon would have resulted, since it would not have been mandatory for supporting the Apollo objectives.

There are other complementary aspects of the findings from the several missions. The 13 successful flights of Rangers, Lunar Orbiters, and Surveyors, plus the 8 trips made by Apollo astronauts, combined to teach us many things we would not have known without the combination. The impact in the crater Alphonsus by Ranger 5, the near polar orbit views of Orbiter 5, the landing of Surveyor 7 near crater Tycho in the rugged highlands, the visit of Apollo 12 to the Surveyor 3 site, the rover excursions by astronauts gathering broader views and samples to couple with point data, the tremendous benefits from returning lunar samples for examination in laboratories here on Earth—these are but indicators, for there is a long list of synergistic benefits from the combined activities.

Our "obedient" spacecraft have done for us some of the things servants might have done for explorers in the past. They have carried our sensors and equipment where we could not go; they have braved the hostile environments of space and other planets; they have never complained of working hours on end, of being turned off forever when their jobs were done, or even of being sacrificed in the name of science. Fortunately, there is nothing wrong with this treatment of inanimate machines. It encourages me to think that endowing us with the capability to build such "creatures" may be a part of God's plan for helping us rise above slavery.

So far in our conquest of space we have discovered no evidence of living beings. If we view the Moon or Mars as territories for future expansion, then we must plan to establish our bases, dig our mines, build our ports, and perform other necessities without help from "the natives." Today, we might think of colonization through transport of those willing to leave Earth and begin new lives elsewhere. Perhaps the development of territories like Oklahoma and Alaska offer parallels for consideration. On the other hand, during the time the hostile extraterrestrial environments of the Moon and the planets are being tamed with environmentally suitable habitats for humans, building and other developments might be best done by machines. Like our spacecraft explorers, they would have no concern for the environment and need no consideration regarding hours or working conditions; they might even be perceived as being "perfectly happy" doing our work for us. Men would be present in limited numbers to apply our special qualities as yet unassembled into automatons, but in roles as supervisors and not as laborers. Almost everyone enjoys being a sidewalk superintendent at times—would it not be fun to watch a lunar base being built by a variety of specialized machines?

What can we expect these machines to be like? There is no simple answer, for they will surely take many forms and play many roles. Perhaps some of them will combine the qualities of man and machine, reminding us of the impressions received by Indians upon viewing the Spanish horses and riders of Cortez, which they thought to be some new form of creature. We are accustomed to seeing a man and a bulldozer at work as a team; it is not hard to imagine such a machine operated by its control and communications unit doing the bidding of a distant master. In the 1960s we spoke several times of an idea for dispatching roving vehicles to perform "Lewis and Clark expeditions" on the Moon while under the supervision of scientists and engineers here on Earth. The concept envisioned tuning in on TV from our armchairs to see what was happening each day, to observe findings in near real time, and to direct future actions. Just think how much more rewarding and exciting that would be than shooting at monsters through the medium of a video game!

I also think it is exciting to consider the challenges of developing special-purpose machines to do the many things that can conceivably be done by robots. Already machines are beginning to do things for us on Earth that they can do better than man. Production facilities are ideal for machines, where routine functions like welding, or assembling parts, painting, or inspecting can be performed precisely by preprogrammed systems. These tasks do not need the higher order of intelligence possessed by man, and the substitution of machines for men in these instances frees minds for more creative ventures. There are many functions to be performed by robots, and as our capabilities to engineer these systems advance, it should be expected that we will improve our machines by giving them more "brainpower."

Already many stories can be told about the uncanny actions of control system processors which had been programmed to perform complex functions. An example comes to mind from Viking that occurred well into the mission, when Mars was approaching full conjunction; that is, when Mars and Earth were about to be on completely opposite sides of the Sun, so that the Viking orbiters not only became blocked from view by Mars during each orbit, but would also be eclipsed by the Sun. The event that triggered Viking Orbiter 1 into an argument with itself was its not being able to see either the Sun or Earth. The problem became apparent to controllers when the orbiter reappeared and its S-band data stream was missing. The X-band link was strong and nominal, and since it made use of the Earth-pointed S-band antenna for some applications, it seemed likely that the orbiter was still oriented properly.

The low-gain link was checked next, for the spacecraft had been programmed to switch to the low-gain mode if the high-gain link were lost for any reason. High-gain data could miss Earth with even slight pointing errors of the orbiter, for example, but low-gain data could be received on Earth even when the spacecraft was well out of alignment (the reason for this kind of automatic emergency procedure). The search sweep quickly located the low-gain data stream, and it was then possible to acquire the engineering information needed to examine the problem.

The problem was traced to the two data processors associated with the orbiter's computer. These processors were programmed to recognize the Sun as an attitude control reference, and they reacted to the loss of the reference by "safing" the spacecraft with an emergency routine that included spacecraft shutdown events, search activities, and the S-band data transmission transfer from high gain to low gain. This procedure was needed to prevent the spacecraft from performing incorrect maneuvers and going out of control if an onboard failure caused the loss of a navigational reference—like the Sun.

When it was known in advance that a natural reference loss such as a Sun occultation was going to occur, the processors had to be told to disregard the loss and inhibit the safing routine. Both processors had been told to disregard the loss of the Sun during solar occultation, but processor B somehow forgot. The result was that B thought something was wrong with A when A did the right thing by disregarding the loss of the Sun. By design, the processor that sensed a problem (or thought it had) became the priority processor. Consequently, when B decided that A was wrong for not reacting with an emergency response to the loss of the Sun, it took charge and shut A off. This story was quickly reconstructed after the low-gain data rate was precisely adjusted by command from Earth, but there was a bit of finesse involved in getting processor B to relinquish control to processor A. The reason was that orbiter processors were designed so that they could not be simply commanded off; engineers actually "fooled" processor B into relinquishing control by deliberately sending it commands which caused it to err; the stored program then automatically switched to A.

We have much to learn about the use of computer processors, as our efforts to date have produced systems far more limited and primitive than many of those possessed by simple creatures all around us. Who can explain the mysteries of the navigation systems used by migratory birds and turtles sufficiently to allow modeling their systems for use in spacecraft? How is it that tiny insects are capable of attitude stabilization within the limits of their weight and volume? How can the blend of chemical, electrical, and mechanical systems present in most creatures teach us to apply similar principles to our machines? Surely we have a long way to go.

From the meager capabilities of the CC&S of Mariner 2, we now find Voyager spacecraft with 27 processors; not only are they performing individual chores, but computers are actually supervising other computers in distant space. This is effective because the supervisory functions require real-time information and rapid responses that humans directing operations from Earth cannot provide. The long time delay between sending commands and receiving acknowledgment from distances of over 500 million miles is an absolute to be reckoned with. Of course, this application of computer supervision places a burden of responsibility on the engineers who have to provide onboard logic and preprogrammed intelligence to send on the mission. The learning ability ascribed to computer applications has been limited, so the necessary background and experience to be used in flight must, for the most part, be anticipated and provided in advance by the humans in charge.

John Casani, Galileo Project manager at JPL, recently told me about engineers sending a load of commands to a Voyager spacecraft on its way to Saturn. Voyager acknowledged receipt of the commands, but replied that it would not execute them as sent, for they would produce unwanted consequences. This seemingly mutinous response was at first alarming. Days later, after detailed study and simulations, a mistake was found in the command series that the spacecraft had properly detected, even though it had been overlooked by its makers. Fortunately, the thoroughness of their preprogramming had exceeded the quality of checkout applied to en route instructions.

However future chapters of history may be written, it is clear that the spacecraft used in our initial exploration of the Moon and planets were effective prodigies—forerunners of a new age. As automatons become more important in our society, the heritage of early electromechanical Mariners, Rangers, and other spacecraft will assume more significance.

Future applications of such sophisticated technologies will remain as reflections of their masters—either good or evil. Thus far automated spacecraft have always served as partners to man, "for the benefit of all mankind." How their descendants serve will depend on the nobility of man.

About the Author

Oran Nicks' lifelong interest in flight began at age 2, when his uncle took him aloft in a barnstorming biplane. He began learning to fly when he was 16, earned an aeronautical engineering degree when he was 19, and obtained a pilot's license in 1945, while serving in the Air Corps. After World War II, he returned to college for a second degree and joined North American Aviation in 1948 as an aeronautical engineer. During his 10 years at North American he was involved in a supersonic cruise missile project (Navaho) and projects related to high-speed flight and rocket launches, conducting research and development testing at three NACA centers and at CalTech's Jet Propulsion Laboratory. He also participated in flight tests of missiles at Edwards Air Force Base and Cape Canaveral.

The launch of the Soviet Sputnik had a profound impact on Nicks' career. He organized a technical group to do space studies at North American and led a team at the Chance Vought Corporation that developed the Scout launch vehicle. He joined NASA in 1960 as Head of Lunar Flight Systems; in 1961 he was named Director of Lunar and Planetary Programs and became responsible for the Mariner missions to Venus and Mars in addition to Ranger, Lunar Orbiter, and Surveyor. During the next 6 years, Nicks directed programs that launched more than 30 U.S. spacecraft toward the Moon and the planets. He then became Deputy Associate Administrator for Space Science and Applications with continuing responsibilities for lunar and planetary programs, and then Acting Associate Administrator for Advanced Research and Technology. In 1970, he became Deputy Director of Langley Research Center, where he was involved in a broad range of activities, including the Viking Mars program and space shuttle technology. His contributions to lunar and planetary programs earned him many awards, including NASA medals for Distinguished Service, Outstanding Leadership, and Exceptional Service.

In 1980, Nicks returned to his first love, aeronautical engineering. He wrote the present volume at Texas A&M University, while doing research, managing an aerodynamic research facility, and directing a Space Research Center.

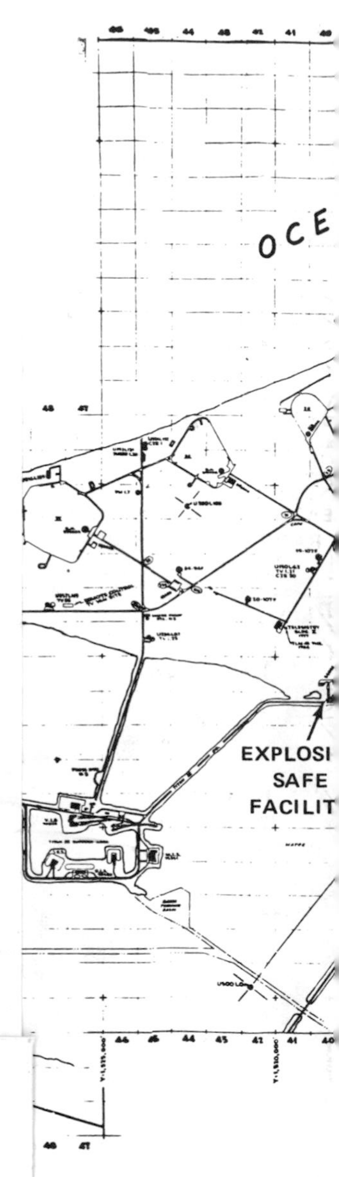

ATLAN

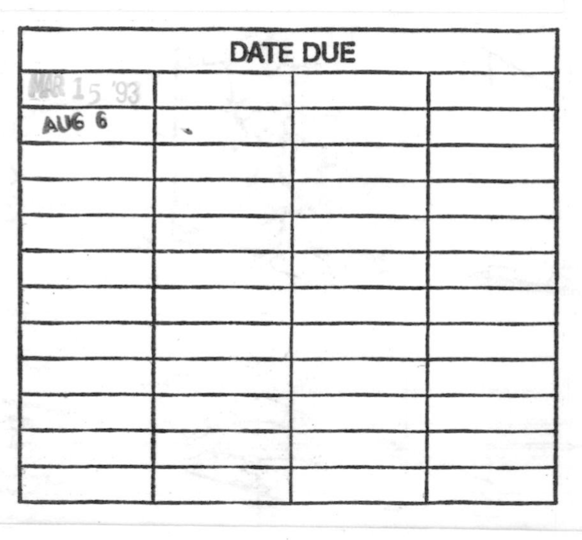

DATE DUE

MAR 15 '93			
AUG 6			